AF358649

Advanced DC-DC Power Converters and Switching Converters

Advanced DC-DC Power Converters and Switching Converters

Editor

Salvatore Musumeci

MDPI • Basel • Beijing • Wuhan • Barcelona • Belgrade • Manchester • Tokyo • Cluj • Tianjin

Editor
Salvatore Musumeci
Politecnico di Torino
Italy

Editorial Office
MDPI
St. Alban-Anlage 66
4052 Basel, Switzerland

This is a reprint of articles from the Special Issue published online in the open access journal *Energies* (ISSN 1996-1073) (available at: https://www.mdpi.com/journal/energies/special_issues/ DC-DC_Converter_Switching_Converters).

For citation purposes, cite each article independently as indicated on the article page online and as indicated below:

LastName, A.A.; LastName, B.B.; LastName, C.C. Article Title. *Journal Name* **Year**, *Volume Number*, Page Range.

ISBN 978-3-0365-0446-9 (Hbk)
ISBN 978-3-0365-0447-6 (PDF)

Contents

About the Editor

Salvatore Musumeci, Ph.D., was born in Giarre, Italy. He received his M.S. in Electrical Engineering and Ph.D. in Electrical Engineering from the University of Catania, Catania, Italy, in 1991 and 1995, respectively. From 1996 to 2001 he worked in the R&D Department of STMicroelectronics (Catania, Italy). From 2001 to 2017, he was involved in several research collaborations with the Department of Electrical Electronic and Systems Engineering, University of Catania (Italy). Since 2018, he has been Assistant Professor of Power Electronics, Electrical Machines, and Drives at the Energy Department of Politecnico di Torino (Italy). He is involved in the Power Electronics Innovation Center (PEIC), a new competence center of Politecnico di Torino (PoliTO) focused on power electronics. His research interests are advanced power devices, switching power converter applications, high-efficiency industrial motors, magnetic materials and their applications, battery management systems, and automotive power electronics applications.

Preface to "Advanced DC-DC Power Converters and Switching Converters"

Nowadays, power electronics is an enabling technology in the energy development scenario. Furthermore, power electronics is strictly linked with several fields of technological growth, such as consumer electronics, IT and communications, electrical networks, utilities, industrial drives and robotics, and transportation and automotive sectors. Moreover, the widespread use of power electronics enables cost savings and minimization of losses in several technology applications required for sustainable economic growth. The topologies of DC–DC power converters and switching converters are under continuous development and deserve special attention to highlight the advantages and disadvantages for use increasingly oriented towards green and sustainable development. DC–DC converter topologies are developed in consideration of higher efficiency, reliable control switching strategies, and fault-tolerant configurations. Several types of switching converter topologies are involved in isolated DC–DC converter and nonisolated DC–DC converter solutions operating in hard-switching and soft-switching conditions. Switching converters have applications in a broad range of areas in both low and high power densities. The articles presented in the Special Issue titled "Advanced DC–DC Power Converters and Switching Converters" consolidate the work on the investigation of the switching converter topology considering the technological advances offered by innovative wide-bandgap devices and performance optimization methods in control strategies used and also in the design of the passive components such as high-frequency isolation transformers. The articles concern switching converter topics such as the following:

- New switching converter topologies for power electronics applications;
- Control and optimization of switching converter circuits;
- Innovative power devices in switching converter applications;
- Advanced DC–DC converters for power supply applications;
- Switching converters in smart grid applications and energy transmission systems;
- Advanced switching converters for renewable energy conversion;
- Advanced DC–DC converters for energy storage systems;
- Switching converters in automotive and traction systems.

From an overview of the articles presented, the issues of the role of converters in the generation of renewable energy and optimization in smart electricity grids together with the problems of recharging batteries for both energy storage systems and electric traction are predominant. As can be seen from the contributions offered, the key role of new semiconductor devices and advanced converter topologies allows a significant contribution to improving energy efficiency. Due to global problems such as the greenhouse effect, energy shortages, and sustainable mobility, a considerable effort is required towards the use of renewable energy and electrical transmission, storage, and implementation systems in the development of livable urban agglomerations and global life quality. Energy conversion via switching converters plays a crucial role in the development of these necessary technological needs. The studies and results presented, while not exhaustive, move in the direction of a further step towards continuous improvement to which we are all called in our research work. Each small research contribution acts in the growth of the quality of life for the well-being of present and especially future generations.

Salvatore Musumeci

Editor

Article

Interleaved High Step-Up DC–DC Converter with Voltage-Lift and Voltage-Stack Techniques for Photovoltaic Systems [†]

Shin-Ju Chen [1], Sung-Pei Yang [1,2,*], Chao-Ming Huang [1] and Yu-Hua Chen [1]

[1] Department of Electrical Engineering, Kun Shan University, Tainan 710303, Taiwan;
    sjchen@mail.ksu.edu.tw (S.-J.C.); h7440@ms21.hinet.net (C.-M.H.); e124667833@gmail.com (Y.-H.C.)
[2] Green Energy Technology Research Center, Kun Shan University, Tainan 710303, Taiwan
* Correspondence: spyang@mail.ksu.edu.tw; Tel.: +886-6-205-1906; Fax: +886-6-205-0298
† This present work is an extension of our paper "High step-up interleaved converter with three-winding coupled inductors and voltage multiplier cells" presented to IEEE ICIT 2019 conference, 13–15 February 2019, Melbourne, Australia.

Received: 13 April 2020; Accepted: 14 May 2020; Published: 16 May 2020

Abstract: A novel interleaved high step-up DC–DC converter applied for applications in photovoltaic systems is proposed in this paper. The proposed configuration is composed of three-winding coupled inductors, voltage multiplier cells and a clamp circuit. The step-up voltage gain is effectively increased, owing to the voltage-stack and voltage-lift techniques using the voltage multiplier cells. The leakage inductor energy is recycled by the clamp circuit to avoid the voltage surge on a power switch. The low-voltage-rated power switches with low on-state resistances and costs can be used to decrease the conduction losses and increase the conversion efficiency when the voltage stresses of power switches for the converter are considerably lower than the high output voltage. The reverse-recovery problems of diodes are mitigated by the leakage inductances of the coupled inductors. Moreover, both the input current ripple and the current stress on each power switch are reduced, owing to the interleaved operation. The operating principle and steady-state analysis of the proposed converter are thoroughly presented herein. A controller network is designed to diminish the effect of the variations of input voltage and output load on the output voltage. Finally, the experimental results for a 1 kW prototype with 28–380 V voltage conversion are shown to demonstrate its effectiveness and performance.

Keywords: interleaved operation; three-winding coupled inductor; high step-up DC–DC converter

---

1. Introduction

Because of the fast exhaustion of fossil fuels and the global warming problem, much research has been developed to cope with green energy sources, such as the fuel cells, photovoltaic power (PV power) or wind power. Generally, a single-phase 220 Vac grid-connected photovoltaic system requires a DC bus voltage of 380–420 V to provide the requirement for a full-bridge DC–AC inverter. Regrettably, the output voltages of individual PV modules are ordinarily lower than 40 V in household applications [1]. Thus, a high step-up DC–DC converter is necessary to serve as a voltage boosting cell between the PV modules and the AC power generation unit [2–4].

For a traditional boost converter, an extreme duty ratio operation has to be realized to obtain a high voltage gain. However, it will result in large current ripples, high conduction losses, reverse-recovery problems for diodes, and electromagnetic interference problems [5]. In addition, the voltage stresses on the power switches and diodes are equal to the high output voltage. Thus, high-voltage-rated MOSFETs with high on-state resistance and diodes with high forward voltage drop should be used, which leads to lower efficiency due to high conduction losses. To proceed, isolated power converters, such as a conventional flyback DC–DC converter, can derive a high voltage gain by adopting a high transformer turns ratio, which results in a large leakage inductance. A large leakage inductance will cause a much higher voltage spike on the power switch and more power dissipations. Consequently, the aforementioned converters are not proper for use in a high step-up voltage gain application.

To overcome the above problems in high voltage gain applications, many high step-up converters have been presented in the literature. Coupled inductors have been adopted to obtain a high voltage gain in the non-isolated converters, because the turns ratio can be served as a control freedom to enlarge the voltage gain [6–10]. Recently, a three-winding coupled inductor has also been applied to a lot of high step-up DC–DC converters to achieve higher voltage gains [11–13]. In [14–19], the switched-inductor and/or switched-capacitor step-up converters are presented to derive a high voltage gain, owing to their simpler structure and operation. A double-duty technique was applied in the high step-up voltage gain applications with two distinct duty ratios for the power switches in [20,21]. The parallel structure on the input side with interleaved operation can be utilized to increase the power level and reduce the input current ripple. The voltage multiplier cells were also applied to the interleaved high step-up converters in [22–24]. The built-in transformer technique for obtaining a high step-up conversion ratio is presented in [25–27]. The interleaved DC–DC converters with three-winding coupled inductors in [28–30] exhibited a high voltage gain and better current sharing performance simultaneously.

An IA novel interleaved high step-up DC–DC converter is proposed in this paper. It contains three-winding coupled inductors, voltage multiplier cells and a clamp circuit. The voltage-stack and voltage-lift techniques are adopted to extend the voltage gain by means of the voltage multiplier cells. The clamp circuit is utilized to recycle the leakage inductor energy and clamp the voltage stress of power switches. The advantages of the proposed high step-up converter are as follows:

(1) By designing a proper turns ratio for the coupled inductors, the high voltage conversion ratio can be obtained whilst operating at an appropriate duty ratio.
(2) The voltage stresses on the power switches are greatly less than the output voltage, so the power switches with lower on-state resistances are utilized to decrease the conduction losses.
(3) The power switches achieve zero-current switching at turn on, and the switching losses can thereby be reduced.
(4) The diode reverse-recovery problem is effectively alleviated by the leakage inductances of the coupled inductors.
(5) The leakage inductor energy can be recycled to suppress the voltage spikes on the power switches.

A prototype of 1 kW was implemented in the laboratory to verify the theoretical analysis and the performance of the proposed interleaved high step-up converter. The remainder of this paper is organized as follows. In Section 2, the circuit description is given, and the operating principle is presented in detail simultaneously. Section 3 shows the steady-state analysis. The performance comparison with existing converters is also presented. The closed-loop controller design is provided in Section 4. Section 5 provides the experimental results of a laboratory prototype. Finally, the conclusion of this paper is given in Section 6.

2. Circuit Description and Operating Principle

Figure 1 shows the circuit topology of the proposed converter. Two three-winding coupled inductors with the same number of turns are included in the proposed converter. The primary, secondary and tertiary windings are denoted by N_1, N_2 and N_3, respectively. The coupling reference is indicated by and *. The primary windings are parallel connected to process the large input current and serve as the filter inductors in the conventional boost converter. The secondary windings are connected in series to constitute the voltage multiplier cell I, which is inserted between the clamp circuit and the output high voltage side to lift the output voltage. The tertiary windings are in series connection to constitute the voltage multiplier cell II, which is stacked on the output capacitor C_1 to enlarge the voltage conversion ratio.

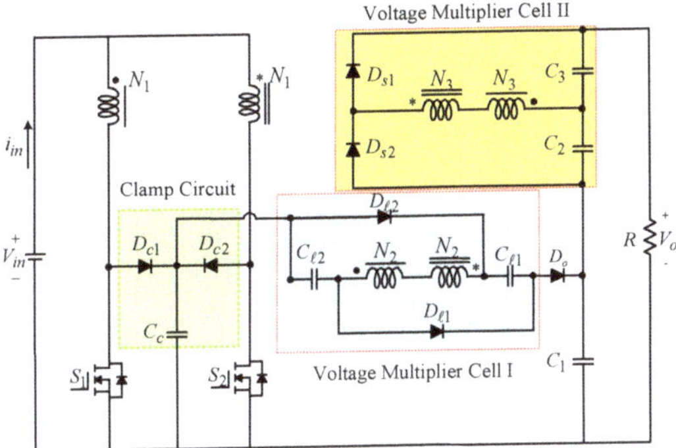

Figure 1. Circuit configuration of the proposed converter.

The coupled inductor is modeled as an ideal transformer with a defined turns ratio, which is in parallel with a magnetizing inductor and in series with a leakage inductor. L_{m1} and L_{m2} represent the magnetizing inductances, while L_{k1} and L_{k2} represent the leakage inductances. Assuming that the number of turns N_3 is equal to N_2. n is defined as the turns ratio with $n = N_2/N_1 = N_3/N_1$. The equivalent circuit of the proposed converter is illustrated in Figure 2, where S_1 and S_2 are the power switches; D_{c1} and D_{c2} are the clamp diodes; C_c is the clamp capacitor; $D_{\ell 1}$ and $D_{\ell 2}$ are the lift diodes; $C_{\ell 1}$ and $C_{\ell 2}$ are the lift capacitors; D_{s1} and D_{s2} are the switched diodes; C_1, C_2 and C_3 are the output capacitors; D_o is the output diode; V_{in} is the input voltage; V_o is the output voltage; and R is the output load.

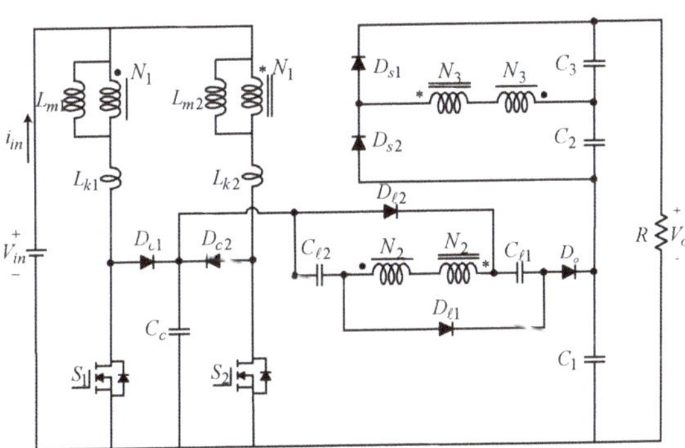

Figure 2. Equivalent circuit of the proposed converter.

The proposed converter operates in continuous conduction mode (CCM). The gate signals of the power switches are interleaved with 180 phase shift, the duty ratios are the same, and they are greater than 0.5. The theoretical waveforms are shown in Figure 3. In CCM operation, the operating mode of the proposed converter can be partitioned into eight stages over one switching period. Figure 4 shows the corresponding circuit models for the eight operating stages.

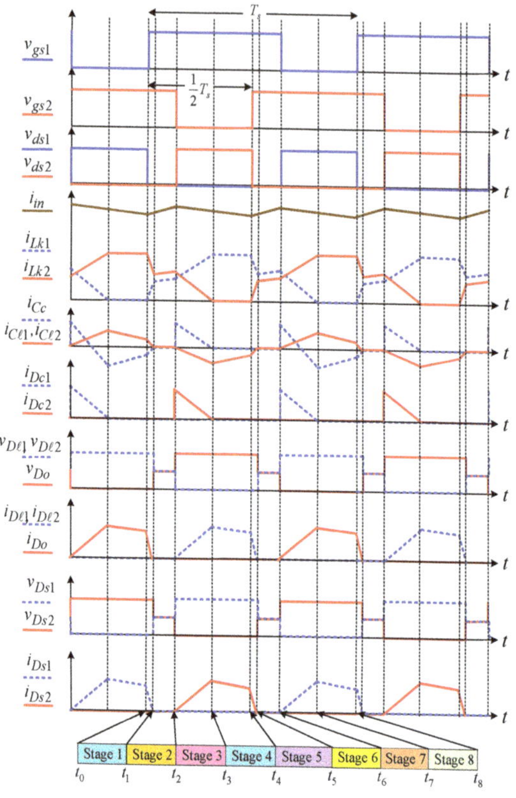

Figure 3. Theoretical waveforms of the proposed converter.

Stage 1 $[t_0 \sim t_1]$: The equivalent circuit of this stage is depicted in Figure 4a. At $t = t_0$, the power switch S_1 starts to turn on with zero-current switching (ZCS) operation, owing to the leakage inductance L_{k1}, and S_2 is still in a turn-on state. The diodes D_{c1}, D_{c2}, $D_{\ell 1}$, $D_{\ell 2}$ and D_{s2} are reversed biased, and D_o as well as D_{s1} are still turned on. The current through L_{k1} increases rapidly from zero, while the currents through the secondary and tertiary windings of the coupled inductors decrease. The current falling rates through D_o and D_{s1} are controlled by the leakage inductances L_{k1} and L_{k2}, such that the diode reverse recovery problem is alleviated. The stored energy in the magnetizing inductor L_{m1} is transferred to the output side via the secondary and tertiary windings of the coupled inductors. The following equations are valid:

$$i_{p2} = -i_{p1} = n(i_{Do} + i_{Ds1}) \tag{1}$$

$$i_{Lk1} = i_{Lm1} + i_{p1} = i_{Lm1} - n(i_{Do} + i_{Ds1}) \tag{2}$$

As the leakage inductor current i_{Lk1} reaches the magnetizing inductor current i_{Lm1}, this stage ends. At the same time, the currents through the diodes D_o and D_{s1} fall to zero, and D_o and D_{s1} are turned off with ZCS operation.

Stage 2 $[t_1 \sim t_2]$: The power switches S_1 and S_2 remain in a turn-on state, and all of the diodes are in a turn-off state. Figure 4b depicts the corresponding operating circuit. The currents through inductors L_{m1}, L_{k1}, L_{m2} and L_{k2} increase linearly because these inductors are charged from the input DC source. The leakage inductor currents are as follows.

$$i_{Lk1}(t) = i_{Lk1}(t_1) + \frac{V_{in}}{L_{m1} + L_{k1}}(t - t_1) \tag{3}$$

$$i_{Lk2}(t) = i_{Lk2}(t_1) + \frac{V_{in}}{L_{m2} + L_{k2}}(t - t_1) \tag{4}$$

This stage ends when S_2 is turned off.

Stage 3 $[t_2 \sim t_3]$: In this stage, the switch S_2 is in a turn-off state, and S_1 keeps conducting. The operating circuit is illustrated in Figure 4c. The clamp capacitor C_c is charged by the current i_{Lk2} via the clamp diode D_{c2}. The leakage inductor energy is released to the capacitor C_c. The current i_{Lk2} decreases linearly. The voltage across the switch S_2 is clamped by the capacitor voltage V_{Cc}. The energy stored in L_{m2} is released to the capacitors $C_{\ell1}$, $C_{\ell2}$ and C_2 via the secondary and tertiary windings of the coupled inductors. The lift capacitors $C_{\ell1}$ and $C_{\ell2}$ are charged by the lift diode currents $i_{D\ell1}$ and $i_{D\ell2}$, respectively. At the same time, the output capacitor C_2 is charged by the current i_{Ds2}. The following equations are valid:

$$i_{p1} = -i_{p2} = ni_{Ds2} + n(i_{D\ell1} + i_{D\ell2}) \tag{5}$$

$$i_{Lk2} = i_{Lm2} - ni_{Ds2} - n(i_{D\ell1} + i_{D\ell2}) \tag{6}$$

The stage finishes as i_{Lk2} falls to zero at $t = t_3$, and the clamp diode D_{c2} becomes reverse-biased under ZCS operation. Thus, there is no reverse recovery loss for D_{c2}.

Stage 4 $[t_3 \sim t_4]$: At the beginning time, the clamp diode D_{c2} is naturally turned off when the leakage inductor energy stored in L_{k2} has fully released to the clamp capacitor C_c. The operating circuit is illustrated in Figure 4d. Magnetizing inductor L_{m2} still transfers its energy to charge $C_{\ell1}$, $C_{\ell2}$ and C_2 via the secondary and tertiary windings of the coupled inductors. The current through the power switch S_1 is the summation of the currents in the magnetizing inductors L_{m1} and L_{m2}. The following equations are held in this stage:

$$i_{Lm2} = n(i_{D\ell1} + i_{D\ell2}) + ni_{Ds2} \tag{7}$$

$$i_{S1} = i_{Lm1} + i_{Lm2} \tag{8}$$

This stage finishes when the turn-on signal is applied to S_2.

Stage 5 $[t_4 \sim t_5]$: In this stage, the operating circuit is depicted in Figure 4e. The switch S_2 turns on at time t_4 under ZCS condition, owing to the leakage inductance L_{k2}, and S_1 is still conducting. The current i_{Lk2} increases rapidly from zero, and the currents in the secondary and tertiary windings of the coupled inductors decrease. The current falling rates through $D_{\ell1}$, $D_{\ell2}$ and D_{s2} are dominated by L_{k1} and L_{k2}, such that the diode reverse recovery problem is mitigated. As the leakage inductor current i_{Lk2} reaches i_{Lm2}, this stage ends at $t = t_5$. At the same time, the currents through $D_{\ell1}$, $D_{\ell2}$ and D_{s2} fall to zero, and these diodes are naturally turned off with ZCS operation.

Stage 6 $[t_5 \sim t_6]$: The switches S_1 and S_2 are conducting in this interval. All of the diodes are in a turn-off state. The operating circuit is depicted in Figure 4f. The operating modes of stages 1 and 6 are similar. At the end of this stage the switch S_1 is turned off.

Stage 7 $[t_6 \sim t_7]$: The switch S_1 is turned off at time t_6. The operating circuit is illustrated in Figure 4g. One part of the leakage inductor energy stored in L_{k1} is released to the clamped capacitor C_c, and another part of the leakage inductor energy is recycled to the output side. The leakage inductor current i_{Lk1} is falling. The input voltage V_{in}, $C_{\ell2}$ and $C_{\ell1}$ are in series connection to transfer energy to the output capacitor C_1 via diodes D_{c1} and D_o, as well as the primary and secondary windings of the coupled inductors, thus extending the voltage on the output capacitor C_1. The stored energy in L_{m1} is delivered to the secondary and tertiary windings of the coupled inductors, such that output capacitor C_3 is charged by the diode current i_{Ds1}, and C_1 is charged by the diode current i_{Do}. As the leakage inductor current i_{Lk1} drops to zero, the diode D_{c1} becomes reverse-biased and turns off at time t_7 under ZCS condition. Thus, there is no reverse recovery loss for D_{c1}. At this moment, this stage ends.

Stage 8 [$t_7 \sim t_8$]: Figure 4h illustrates the operating circuit. At the beginning time, the leakage inductor energy stored in L_{k1} has completely released. Magnetizing inductor L_{m1} still transfers energy to the capacitors C_1 and C_3 via the secondary and tertiary windings of the coupled inductors. The capacitors C_c, $C_{\ell1}$, $C_{\ell2}$ and the secondary windings are connected in series to transfer their energy to the output capacitor C_1. The current in the switch S_2 is the summation of the currents i_{Lm1} and i_{Lm2}. The switch S_1 is turned on at the end of this stage. Then, a new switching period begins to start.

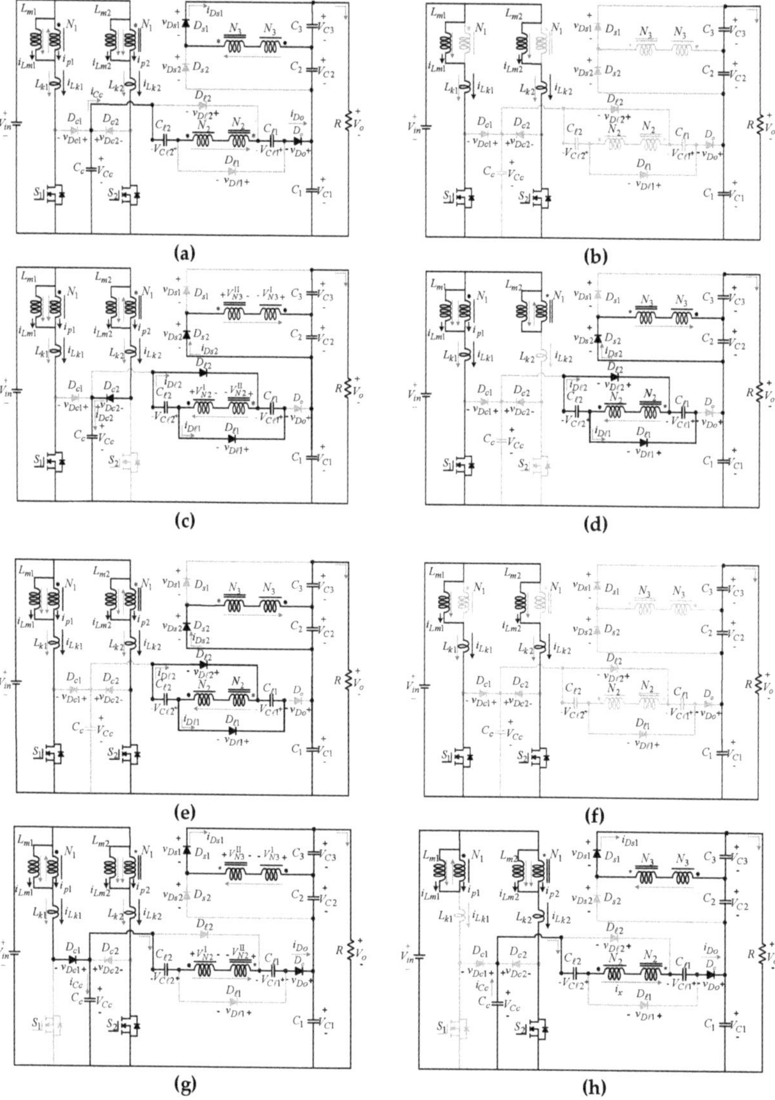

Figure 4. Operating stages of the proposed converter. (**a**) Stage 1, (**b**) Stage 2, (**c**) Stage 3, (**d**) Stage 4, (**e**) Stage 5, (**f**) Stage 6, (**g**) Stage 7, (**h**) Stage 8.

3. Steady-State Analysis

3.1. Voltage Gain Derivation

To briefly describe the voltage gain derivation, the following assumptions are used:

(1) All of the semiconductors are regarded as ideal. The on-state resistance of the switches and the forward voltage drop of the diodes are ignored.
(2) The leakage inductances are neglected.
(3) The magnetizing inductances of the coupled inductors are regarded as the same; that is, $L_{m1} = L_{m2} = L_m$.

All of the capacitors are large enough. As a result, the voltages across them are considered constant during one switching period. Based on the volt-second balance principle of the magnetizing inductance L_{m1}, the voltage on the clamp capacitor C_c can be derived as

$$V_{Cc} = \frac{1}{1-D} V_{in} \tag{9}$$

where D is the operating duty ratio. The result in Equation (9) is identical to the output voltage of a conventional boost converter.

Let the voltages across the secondary and tertiary windings of the coupled inductors be denoted by V_{N2}^{I} and V_{N3}^{I}, V_{N2}^{II} and V_{N3}^{II}, respectively. According to Kirchhoff's Voltage Low (KVL), the voltages across the lift capacitors $C_{\ell 1}$ and $C_{\ell 2}$ can be calculated from stage 3 as

$$V_{C\ell 1} = V_{C\ell 2} = V_{N2}^{I} - V_{N2}^{II} = nV_{in} - n(V_{in} - V_{Cc}) = nV_{Cc} \tag{10}$$

Moreover, it also yields

$$V_{C2} = V_{N3}^{I} - V_{N3}^{II} = nV_{in} - n(V_{in} - V_{Cc}) = nV_{Cc} \tag{11}$$

Substituting Equation (9) into Equations (10) and (11), the capacitor voltages are rewritten as

$$V_{C\ell 1} = V_{C\ell 2} = \frac{n}{1-D} V_{in} \tag{12}$$

$$V_{C2} = \frac{n}{1-D} V_{in} \tag{13}$$

By applying KVL in stage 7, the voltage V_{C3} across the output capacitor C_3 can be derived as

$$V_{C3} = V_{N3}^{II} - V_{N3}^{I} = nV_{Cc} = \frac{n}{1-D} V_{in} \tag{14}$$

Moreover, the voltage across the output capacitor C_1 is derived as

$$V_{C1} = V_{N2}^{II} - V_{N2}^{I} + V_{Cc} + V_{C\ell 1} + V_{C\ell 2} = \frac{3n+1}{1-D} V_{in} \tag{15}$$

According to (13)–(15), the output voltage can be obtained as follows:

$$V_o = V_{C1} + V_{C2} + V_{C3} = \frac{5n+1}{1-D} V_{in} \tag{16}$$

Hence, we have the ideal voltage gain M of the proposed converter as

$$M = \frac{V_o}{V_{in}} = \frac{5n+1}{1-D} \tag{17}$$

The plot of voltage gain M versus turns ratio n and duty ratio D is drawn in Figure 5. It shows that the turns ratio has a significant impact on the step-up voltage gain. In addition, the high voltage gain can be achieved without any extreme duty ratio or high turns ratio in the proposed converter. When the duty ratio is merely 0.6 and turns ratio $n = 1$, the voltage gain is calculated as 15.

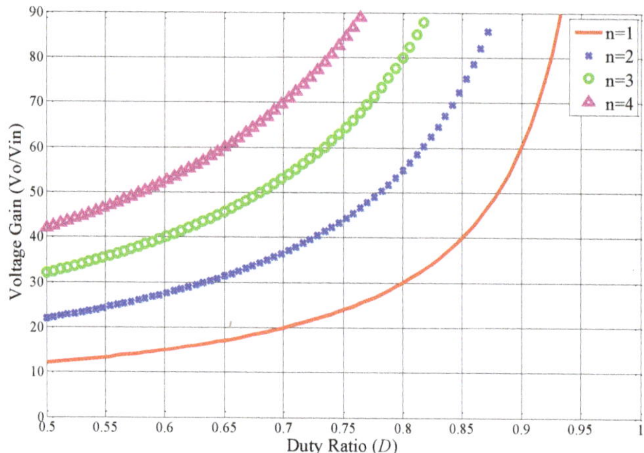

Figure 5. Voltage gain curve versus duty ratio with different turns ratio.

3.2. Voltage Stresses on Semiconductor Devices

The steady-state analysis reveals that the voltage on the power switches and the clamp diodes during their off-state are all equal to the voltage on the clamp capacitor. From Equations (9) and (17), the voltage stresses are given by

$$V_{S1} = V_{S2} = V_{Dc1} = V_{Dc2} = V_{Cc} = \frac{1}{1-D}V_{in} = \frac{1}{5n+1}V_o \tag{18}$$

Moreover, the voltage stress on the switching diode D_{s1} can be derived as

$$V_{Ds1} = V_{C2} + V_{C3} = \frac{2n}{1-D}V_{in} = \frac{2n}{5n+1}V_o \tag{19}$$

The voltage stress on the output diode D_o is given by

$$V_{Do} = V_{C1} - V_{C\ell1} - V_{Cc} = \frac{2n}{1-D}V_{in} = \frac{2n}{5n+1}V_o \tag{20}$$

Similarly, the voltage stresses on the diodes D_{s2}, $D_{\ell1}$ and $D_{\ell2}$ can be derived as

$$V_{Ds2} = V_{D\ell1} = V_{D\ell2} = \frac{2n}{1-D}V_{in} = \frac{2n}{5n+1}V_o \tag{21}$$

From Equations (18)–(21), the relationship between the normalized voltage stresses on semiconductor devices and the turns ratio of the coupled inductors is shown in Figure 6.

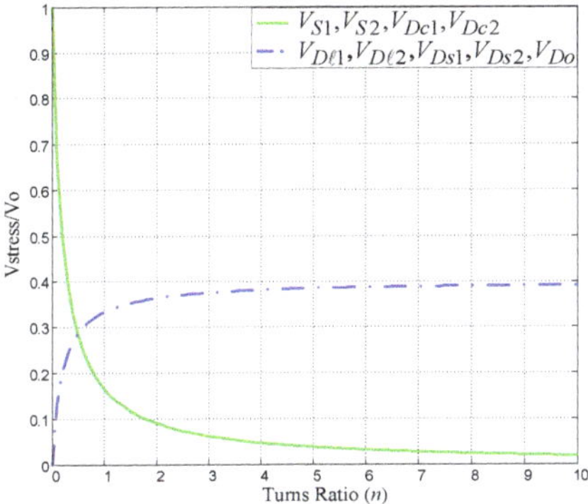

Figure 6. Normalized voltage stresses on semiconductor devices.

As the turns ratio increases, the voltage stresses on S_1, S_2, D_{c1} and D_{c2} decrease, and the voltage stresses on the other diodes become large. It is worth noting that the voltage stresses are lower than the output voltage. As a result, power MOSFETs with low $R_{ds(ON)}$ and diodes with low forward voltage drop can be employed to reduce the on-state losses and improve the conversion efficiency.

3.3. Design Considerations

3.3.1. Design of Coupled Inductors

The turns ratio of the coupled inductors is designed from Equation (17). Once the duty ratio has been selected, the turns ratio n can be properly designed by

$$n = \frac{N_3}{N_1} = \frac{N_2}{N_1} = \frac{(1-D)V_o}{5V_{in}} - \frac{1}{5} \tag{22}$$

Once the turns ratio of the coupled inductor is obtained, the magnetizing inductance can be determined from the CCM operation mode and an acceptable current ripple. The current ripple on the magnetizing inductor is identical, and given by

$$\Delta i_{Lm} = \frac{V_{in}D}{L_m f_s} \tag{23}$$

where f_s is the switching frequency. The average magnetizing current can be derived as

$$I_{Lm} = \frac{P_o}{2V_{in}} = \frac{V_o^2}{2V_{in}R} \tag{24}$$

where P_o is the output power. For CCM operation, the following condition holds:

$$I_{Lm} - \frac{1}{2}\Delta i_{Lm} > 0 \tag{25}$$

Substituting Equations (23) and (24) into (25), the condition of magnetizing inductance for CCM operation is expressed as

$$L_m > \frac{V_{in}^2 D}{P_o f_s} = \frac{D(1-D)^2 V_o^2}{(5n+1)^2 P_o f_s} = \frac{D(1-D)^2 R}{(5n+1)^2 f_s} \tag{26}$$

3.3.2. Design of Capacitors

The capacitors are determined to limit their voltage ripples to within an acceptable range. The output capacitor C_1 is discharged by the average load current I_o from Stage 2 to Stage 6. Thus, its voltage ripple can be derived as

$$\Delta V_{C1} = \frac{DI_o}{C_1 f_s} \tag{27}$$

Substituting Equations (15) and (17) into (27), the required capacitance is calculated as

$$C_1 = \frac{(5n+1)D}{(3n+1)Rf_s(\Delta V_{C1}/V_{C1})} \tag{28}$$

which is expressed by the specified voltage ripple on the output capacitor C_1. Similarly, one can obtain the design of the following capacitors in terms of their own specified voltage ripples:

$$C_2 = C_3 = \frac{(5n+1)D}{nRf_s(\Delta V_{Ci}/V_{Ci})}, \ i = 2,\ 3 \tag{29}$$

$$C_c = \frac{5n+1}{Rf_s(\Delta V_{Cc}/V_{Cc})} \tag{30}$$

3.4. Performance Comparison

Table 1 shows the performance comparison between the proposed converter and some interleaved high step-up converters published in [28–30], including voltage gain, voltage stress on switches, maximum diode voltage stress and the quantities of the devices. In these comparative converters, three-winding coupled inductors are employed to achieve high step-up voltage gain. Figure 7 shows the voltage gain comparison with turns ratio $n = 1$. As can be seen, the proposed converter has the highest voltage gain. In addition, it also has the lowest voltage stresses on the switches and diodes. The voltage stresses on the semiconductor devices are lower than the high output voltage, which results in the use of switches with low on-resistance and diodes with low forward voltage drop to reduce the conduction losses and improve the conversion efficiency. As a result, it is clear that the proposed converter is very suitable for applications requiring high efficiency and a high step-up voltage conversion ratio.

$$C_{\ell 1} = C_{\ell 2} = \frac{5n+1}{nRf_s(\Delta V_{C\ell j}/V_{C\ell j})}, j = 1,2 \tag{31}$$

Table 1. Performance comparison of characteristics.

Converter	Converter in [28]	Converter in [29]	Converter in [30]	Proposed Converter
Voltage gain	$\frac{2n+2}{1-D}$	$\frac{2n+2}{1-D}$	$\frac{3n+1}{1-D}$	$\frac{5n+1}{1-D}$
Voltage stress on switches	$\frac{V_o}{2n+2}$	$\frac{V_o}{2n+2}$	$\frac{V_o}{3n+1}$	$\frac{V_o}{5n+1}$
Maximum diode voltage stress	$\frac{(2n+1)V_o}{2n+2}$	$\frac{(2n+1)V_o}{2n+2}$	$\frac{2nV_o}{3n+1}$	$\frac{2nV_o}{5n+1}$
Quantities of switches	2	2	2	2
Quantities of diodes	6	6	8	7
Quantities of capacitors	5	5	7	6
Quantities of coupled inductors	2	2	2	2
Maximal efficiency at output power	95.8% at 500W	97.2% at 400W	97% at 524W	98% at 100W

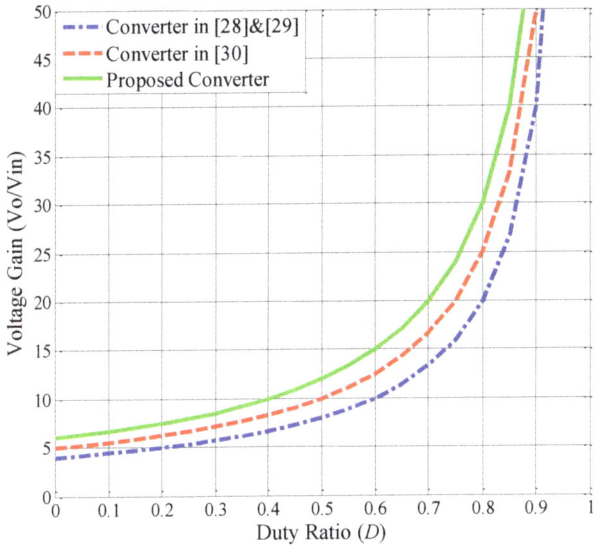

Figure 7. Voltage gain comparison with turns ratio $n = 1$.

4. Controller Design

For the purposes of the output voltage regulation, regardless of the variations of input voltage and output load, the voltage-mode feedback control system was built as shown in Figure 8. Blocks $C(s)$ and PWM represent the controller and pulse-width modulator, respectively. Block $P(s)$ denotes the converter power stage. Block K denotes the sensor gain.

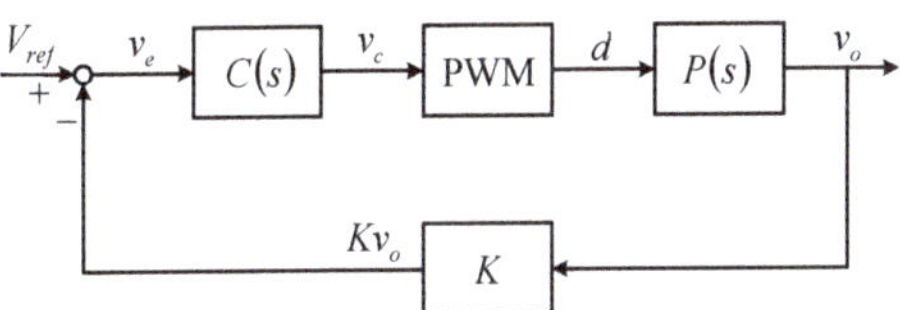

Figure 8. Control system block diagram.

A small-signal model was investigated through the frequency response with experimental measurements for the prototype converter. The electrical specifications and component parameters of the prototype converter are shown in Table 2. The experimental frequency response at the operating point of half load was measured by an NF FRA5012 frequency response analyzer. The Bode plot of the measured transfer function from control to scaled output voltage ($\widetilde{v}_c$ to $K\widetilde{v}_o$) is shown in Figure 9 with red curves. The corresponding transfer function can be obtained by the curve-fitting method, and it is given by

$$G(s) = \frac{K\widetilde{v}_o(s)}{\widetilde{v}_c(s)} = \frac{138.3(s - 45000)}{(s + 700)(s + 7000)} \tag{32}$$

Table 2. Electrical specifications and parameters of the prototype converter.

Parameter/Description	Specification/Value
Input voltage V_{in}	28 V
Output voltage V_o	380 V
Rated output power P_o	1000 W
Switching frequency f_s	50 kHz
Magnetizing inductance L_m	69 μH
Leakage inductance L_k	0.7 μH
Turns ratio n	1
Power switches S_1 and S_2	IRFP4668
Diodes $D_{C1}, D_{C2}, D_{\ell1}, D_{\ell2}, D_{S1}, D_{S2}$ and D_o	60CPQ150
Clamp capacitor C_C	147 μF
Lift capacitors $C_{\ell1}$ and $C_{\ell2}$	147 μF
Output capacitors C_1, C_2 and C_3	120 μF

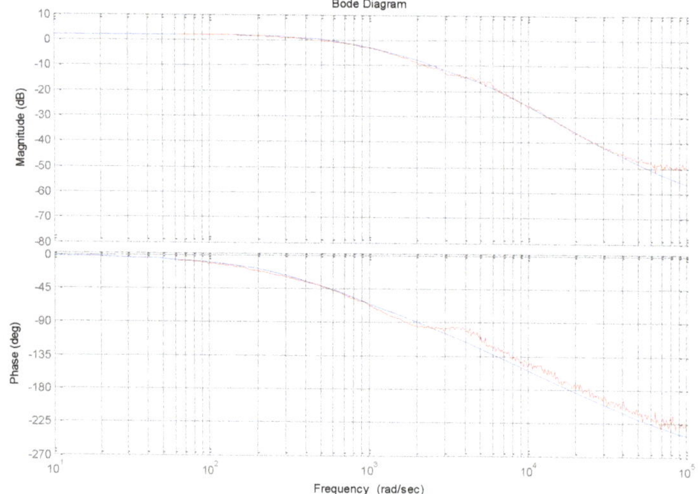

Figure 9. Comparison between measured (red) and curve-fitting (blue).

The Bode plot of the curve-fitting transfer function in Equation (32), together with the measured results, is shown in Figure 9. Comparing the magnitude and phase curves, it can be seen that the curves agree well with each other. Thus, the curve-fitting transfer function expressed in Equation (32) can be used for the controller design.

Based on the *K*-factor method [31], a type III controller [32] with three-pole and two-zero was designed for the closed-loop control system. One of the poles of the controller was located at the origin to achieve the zero steady-state error, while the other two poles were positioned below the desired crossover frequency to attenuate the switching noises in the feedback loop. In addition, the zeros and gain of the controller were adjusted to achieve the desired phase margin at the crossover frequency. The controller transfer function was designed as

$$C(s) = 3.3 \times 10^6 \frac{(s + 2659)(s + 2673)}{s(s + 1.49 \times 10^4)^2} \tag{33}$$

The controller was implemented by the operational amplifier circuit, as shown in Figure 10, and its transfer function is given by

$$\frac{\widetilde{v}_c(s)}{K\widetilde{v}_o(s)} = -\frac{R_1 + R_3}{R_1 R_3 C_2} \frac{\left(s + \frac{1}{R_2 C_1}\right)\left(s + \frac{1}{(R_1 + R_3)C_3}\right)}{s\left(s + \frac{1}{R_2 C_1 C_2/(C_1 + C_2)}\right)\left(s + \frac{1}{R_3 C_3}\right)} \tag{34}$$

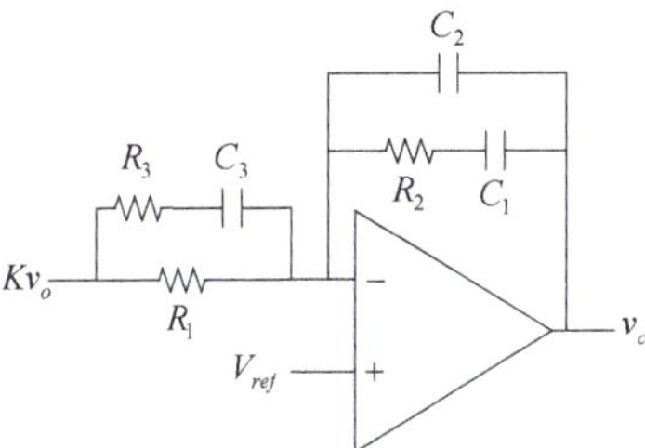

Figure 10. Controller circuit.

The Bode plots of the plant $G(s)$, the controller $C(s)$ and the loop gain $\text{Tol}(s) = G(s)C(s)$ are shown in Figure 11. As a result, a crossover frequency of 1 kHz and a phase margin of 45° were achieved for the output voltage controlled system.

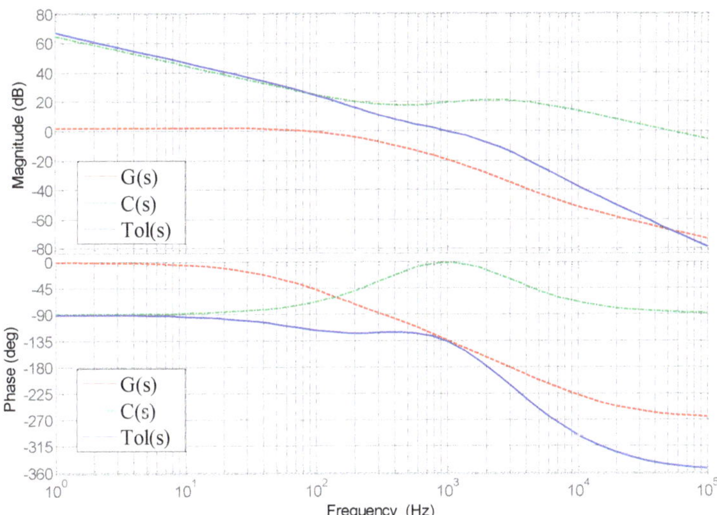

Figure 11. Bode plots of plant, controller and loop gain.

5. Experimental Verification

An experimental prototype with maximal output power 1 kW was implemented and tested to verify the performance of the proposed converter. Table 2 shows the components and parameters of the prototype converter [33]. Figures 12–16 show the simulated results using IsSpice software and the experimental results under full load 1 kW condition, as described below.

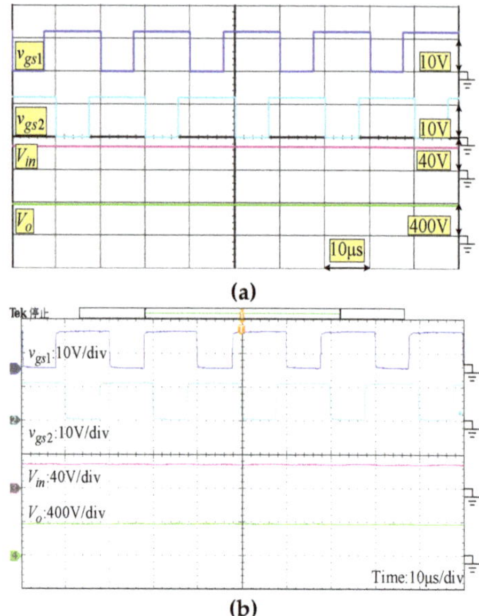

(a)

(b)

Figure 12. Waveforms of v_{gs1}, v_{gs2}, V_{in} and V_o. (**a**) Simulated results. (**b**) Experimental results.

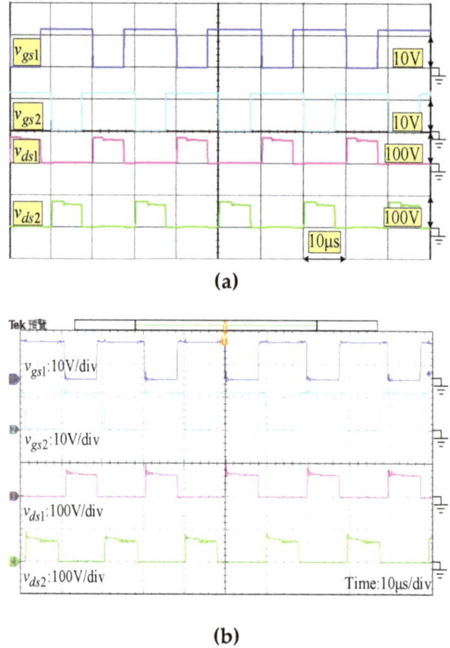

(a)

(b)

Figure 13. Waveforms of v_{gs1}, v_{gs2}, v_{ds1} and v_{ds2}. (**a**) Simulated results. (**b**) Experimental results.

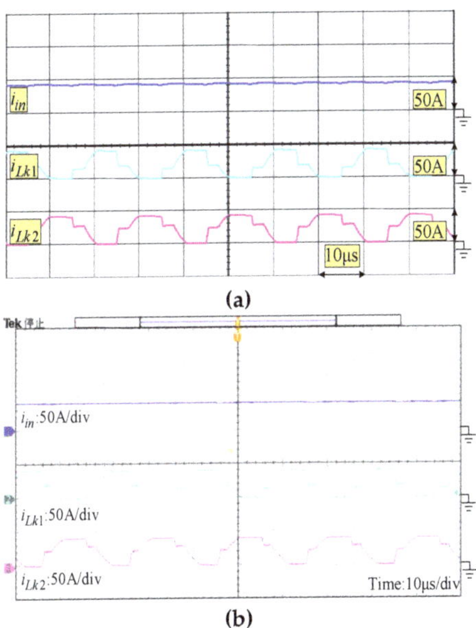

(a)

(b)

Figure 14. Waveforms of i_{in}, i_{Lk1} and i_{Lk2}. (**a**) Simulated results. (**b**) Experimental results.

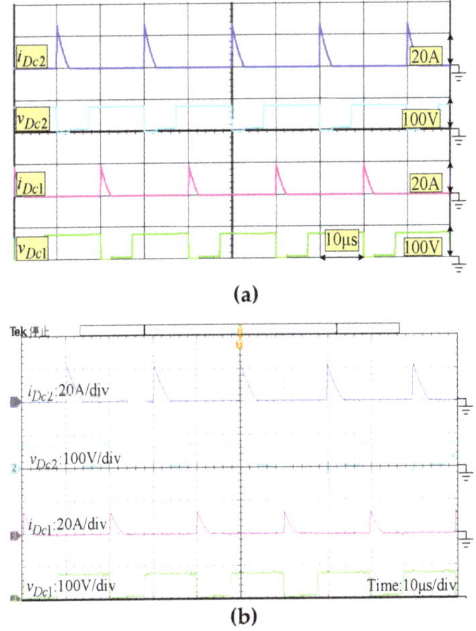

(a)

(b)

Figure 15. Waveforms of i_{Dc1}, v_{Dc1}, i_{Dc2} and v_{Dc2}. (**a**) Simulated results. (**b**) Experimental results.

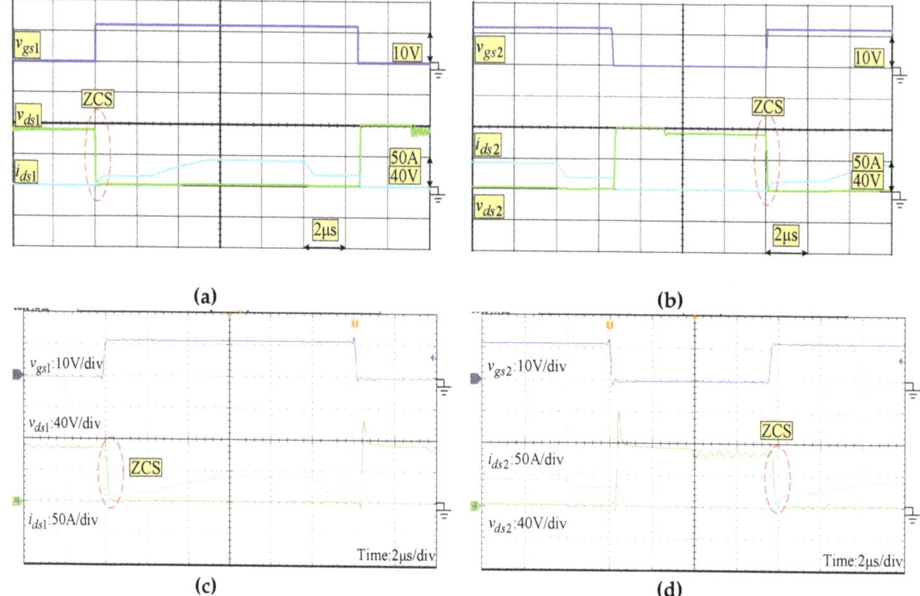

Figure 16. Waveforms of the ZCS turn-on for switches S_1 and S_2. (**a**) and (**b**) simulated results. (**c**) and (**d**) experimental results.

Figure 12 shows the waveforms of V_{in}, V_o, and the gate signals v_{gs1} and v_{gs2}, for the switches S_1 and S_2 with interleaved operation. It can be seen that the high voltage gain was over 13 times; however, the duty ratios of the switches were not extremely large.

Figure 13 illustrates the gate signals and the drain-source voltage waveforms v_{ds1} and v_{ds2} for the switches S_1 and S_2. It was observed that the voltage stress on S_1 and S_2 was only about 63 V, which is $V_o/6$. The switch voltage stress was much lower than the output voltage. This result meets with that of the steady-state analysis in Equation (18). Therefore, the power switch with a low voltage rating and low on-resistance can be chosen to reduce the conduction losses.

Figure 14 represents the input current i_{in} and the leakage inductor currents i_{Lk1} and i_{Lk2}. Since the input current i_{in} is equal to i_{Lk1} plus i_{Lk2}, one can see that the interleaved operation helps the ripple current cancellation. Consequently, the input current ripple is really small. The ripple current reduction is helpful for the lifetime of green energy sources. Moreover, a good input current sharing capability can be observed by the leakage inductor currents for the two phases of the proposed converter.

Figure 15 exhibits the currents and voltage waveforms on the clamped diodes D_{c1} and D_{c2}. One can see that the voltage stress on the diodes is about 63 V, which is only one-sixth of the output voltage. The experimental results show good agreement with the theoretical result in (18). In addition, as can be seen, the currents i_{Dc1} and i_{Dc2} fell to zero, and then the considered diodes turned off with the ZCS operation, which is consistent with the operating analysis in stages 3 and 7. Thus, there are no reverse-recovery losses for the clamped diodes D_{c1} and D_{c2}.

In Figure 16, the simulated and experimental waveforms of the voltages and the currents on the switches S_1 and S_2 are illustrated. It can be seen that the power switches can achieve ZCS turn-on operation. The switching losses are reduced accordingly for high efficiency.

Figure 17 shows the dynamic response of the output voltage under the load variation between 200W and 1000W using a dc electronic load. The dynamic response of the output voltage under the input voltage varying from 28 V to 32 V, and vice versa, is shown in Figure 18. As shown in the figures, the output voltage is insensitive to the load change and input voltage variation. It means that the well dynamic performance of the output voltage regulation can be provided with the closed-loop controller design.

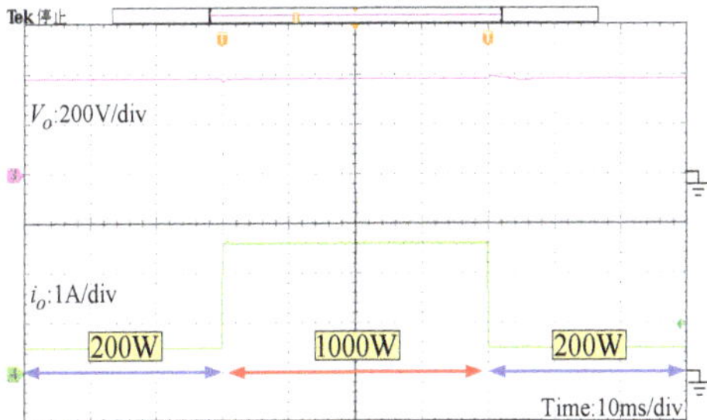

Figure 17. Dynamic response of output voltage under step load variation.

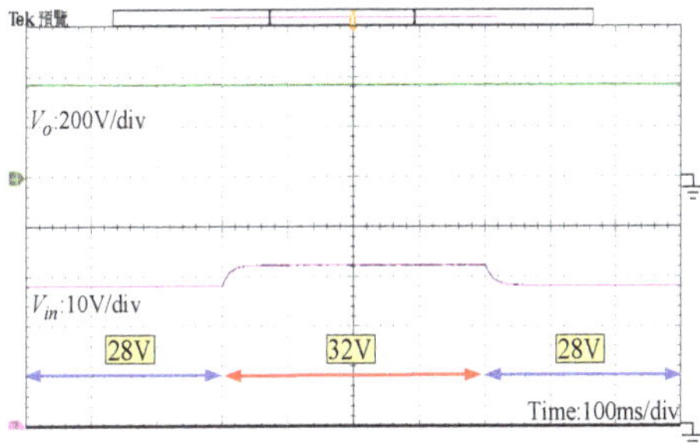

Figure 18. Dynamic response of output voltage under input voltage variation.

Figure 19 represents the conversion efficiency of the prototype converter under various output powers. A high precision power analyzer (HIOKI 3390) was employed to measure the power conversion efficiency, which is the ratio of the measured output power over the measured input power, Pout/Pin. The measured maximum conversion efficiency was up to 98%, which was obtained at the output power of 100 W. Moreover, the conversion efficiency was 91.08% at the full-load condition. At higher output power, the on-state conduction losses of switching devices are high. Then, the efficiency decreases. The photograph of the laboratory prototype is illustrated in Figure 20.

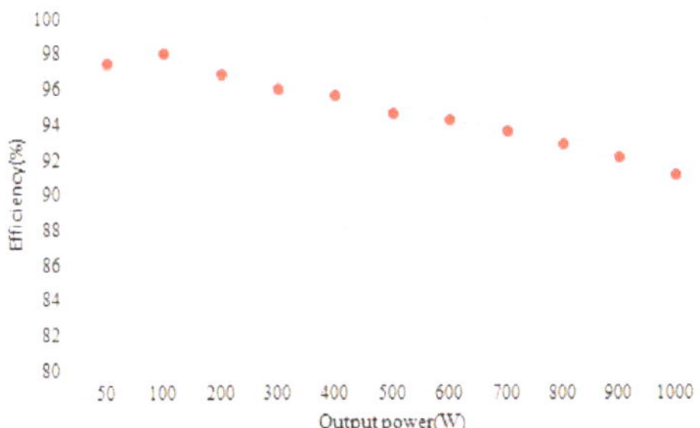

Figure 19. Power conversion efficiency.

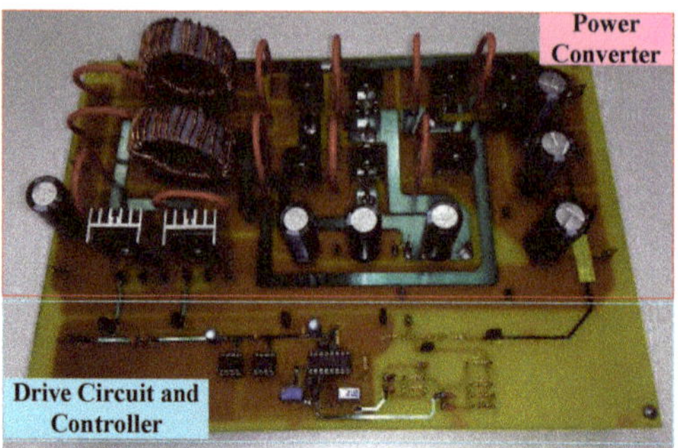

Figure 20. Photograph of the laboratory prototype.

6. Conclusions

The three-winding coupled inductors and voltage multiplier cells, and the voltage-lift and voltage-stack techniques were utilized to create a novel high step-up DC–DC converter configuration, which is suitable for applications in PV generation systems. The proposed high step-up converter gets high voltage gain conversion with proper duty ratio operation and low voltage stresses on the switches and diodes. Switches with smaller on-resistance and diodes with lower forward voltage drop can thereby be used to reduce the conduction losses. The interleaved operation reduces the input current ripple. Moreover, the diode reverse-recovery loss is alleviated due to the leakage inductances of the coupled inductors. The leakage inductor energy is absorbed and recycled to improve efficiency. This paper presented the operating principle and steady-state analysis of the proposed converter. The closed-loop controller is also well designed for the output voltage regulation, regardless of the variations in the input voltage or output load. Finally, a 1 kW laboratory prototype was tested to verify the performance and the presented analysis. The experimental results showed that the proposed converter is suitable for high efficiency and high voltage gain in DC–DC conversion.

Author Contributions: S.-J.C. and S.-P.Y.: analysis and design. Y.-H.C.: experiment. C.-M.H.: supervision and inspection. S.-J.C. and S.-P.Y.: writing and editing. All authors have read and agreed to the published version of the manuscript.

Funding: This work was financially supported by the Green Energy Technology Research Center from The Featured Areas Research Center Program within the framework of the Higher Education Sprout Project by the Ministry of Education (MOE) in Taiwan, and the Ministry of Science and Technology, Taiwan, under grant numbers MOST 108-2221-E-168-004.

Conflicts of Interest: The authors declare that there is no conflict of interest.

References

1. Ajami, A.; Ardi, H.; Farakhor, A. A novel high step-up DC-DC converter based on integrating coupled inductor and switched-capacitor techniques for renewable energy applications. *IEEE Trans. Power Electron.* **2015**, *30*, 4255–4263. [CrossRef]
2. Yang, B.; Li, W.; Zhao, Y.; He, X. Design and analysis of a grid connected PV power system. *IEEE Trans. Power Electron.* **2010**, *25*, 992–1000. [CrossRef]
3. Carrasco, J.M.; Franquelo, L.G.; Bialasiewicz, J.T.; Galvan, E.; PortilloGuisado, R.C.; Prats, M.A.M.; Leon, J.I.; Moreno-Alfonso, N. Power-electronic systems for the grid integration of renewable energy sources: A survey. *IEEE Trans. Ind. Electron.* **2006**, *53*, 1002–1016. [CrossRef]
4. Andrade, A.M.; Schuch, L.; Martins, M.L.S. Analysis and design of high-efficiency hybrid high step-up DC–DC converter for distributed PV generation systems. *IEEE Trans. Ind. Electron.* **2019**, *66*, 3860–3868. [CrossRef]
5. Tofoli, F.L.; Pereira, D.C.; Paula, W.J.; Junior, D.S.O. Survey on non-isolated high-voltage step-up DC–DC topologies based on the boost converter. *IET Power Electron.* **2015**, *8*, 2044–2057. [CrossRef]

6. Wong, Y.S.; Chen, J.F.; Liu, K.B.; Hsieh, Y.P. A novel high step-up DC-DC converter with coupled inductor and switched clamp capacitor techniques for photovoltaic systems. *Energies* **2017**, *10*, 378–394. [CrossRef]
7. Hu, X.; Gong, C. A high voltage gain DC-DC converter integrating coupled-inductor and diode-capacitor techniques. *IEEE Trans. Power Electron.* **2014**, *29*, 789–800.
8. Farakhor, A.; Abapour, M.; Sabahi, M.; Farkoush, S.G.; Oh, S.R.; Rhee, S.B. A study on an improved three-winding coupled inductor based dc/dc boost converter with continuous input current. *Energies* **2020**, *13*, 1780. [CrossRef]
9. Tseng, K.C.; Lin, J.T.; Huang, C.C. High step-up converter with three-winding coupled inductor for fuel cell energy source applications. *IEEE Trans. Power Electron.* **2015**, *30*, 574–581. [CrossRef]
10. Yu, D.; Yang, J.; Xu, R.; Xia, Z.; Iu, H.H.; Fernando, T. A family of module-integrated high step-up converters with dual coupled inductors. *IEEE Access* **2018**, *6*, 16256–16266. [CrossRef]
11. Azizkandi, M.E.; Sedaghati, F.; Shayeghi, H.; Blaabjerg, F. Two- and three-winding coupled-inductor-based high step-up DC–DC converters for sustainable energy applications. *IET Power Electron.* **2020**, *13*, 144–156.
12. Gao, Y.; Liu, H.; Ai, J. Novel high step-up DC–DC converter with three-winding-coupled-inductors and its derivatives for a distributed generation system. *Energies* **2018**, *11*, 3428. [CrossRef]
13. Azizkandi, M.E.; Sedaghati, F.; Shayeghi, H.; Blaabjerg, F. A high voltage gain DC–DC converter based on three winding coupled inductor and voltage multiplier cell. *IEEE Tran. Power Electron.* **2020**, *35*, 4558–4567. [CrossRef]
14. Wei, C.L.; Shih, M.H. Design of a switched-capacitor DC-DC converter with a wide input voltage range. *IEEE Trans. Circuits Syst.* **2013**, *60*, 1648–1656. [CrossRef]
15. Tang, Y.; Fu, D.; Wang, T.; Xu, Z. Hybrid switched-inductor converters for high step-up conversion. *IEEE Trans. Ind. Electron.* **2015**, *62*, 1480–1490. [CrossRef]
16. Axelrod, B.; Berkovich, Y.; Ioinovici, A. Switched-capacitor/switched-inductor structures for getting transformerless hybrid DC–DC PWM converters. *IEEE Trans. Circuits Syst.* **2008**, *55*, 687–696. [CrossRef]
17. Salvador, M.A.; Lazzarin, T.B.; Coelho, R.F. High step-up DC-DC converter with active switched-inductor and passive switched-capacitor networks. *IEEE Trans. Ind. Electron.* **2018**, *65*, 5644–5654. [CrossRef]
18. Gu, Y.; Chen, Y.; Zhang, B.; Qiu, D.; Xie, F. High step-up DC–DC converter with active switched LC-network for photovoltaic systems. *IEEE Trans. Energy Convers.* **2019**, *34*, 321–329. [CrossRef]
19. Duong, T.D.; Nguyen, M.K.; Tran, T.T.; Lim, Y.C.; Choi, J.H. Transformerless high step-up DC-DC converters with switched-capacitor network. *Electronics* **2019**, *8*, 1420. [CrossRef]
20. Lakshmi, M.; Hemamalini, S. Nonisolated high gain DC-DC converter for DC microgrids. *IEEE Trans. Ind. Electron.* **2018**, *65*, 1205–1212. [CrossRef]
21. Bhaskar, M.S.; Meraj, M.; Iqbal, A.; Padamanaban, S.; Maroti, P.K.; Alammari, R. High gain transformer-less double-duty-triple-mode DC/DC converter for DC microgrid. *IEEE Access* **2019**, *7*, 36353–36370. [CrossRef]
22. Tseng, K.C.; Chen, C.T.; Cheng, C.A. A high-efficiency high step-up interleaved converter with a voltage multiplier for electric vehicle power management applications. *J. Power Electron.* **2016**, *16*, 414–424. [CrossRef]
23. Zhou, W.B.; Zhu, X.; Luo, Q.M.; Chen, S. Interleaved non-isolated high step-up dc/dc converter based on the diode-capacitor multiplier. *IET Power Electron.* **2014**, *7*, 390–397. [CrossRef]
24. Li, W.; Zhao, Y.; Deng, Y.; He, X. Interleaved converter with voltage multiplier cell for high step-up and high-efficiency conversion. *IEEE Trans. Power Electron.* **2010**, *25*, 2397–2408. [CrossRef]
25. Li, W.; Xiang, X.; Li, C.; Li, W.; He, X. Interleaved high step-up ZVT converter with built-in transformer voltage doubler cell for distributed PV generation system. *IEEE Trans. Power Electron.* **2013**, *28*, 300–313. [CrossRef]
26. Li, W.; Xiang, X.; Hu, Y.; He, X. High step-up interleaved converter with built-in transformer voltage multiplier cells for sustainable energy applications. *IEEE Trans. Power Electron.* **2014**, *29*, 2829–2836. [CrossRef]
27. Nouri, T.; Vosoughi, N.; Hosseini, S.H.; Babaei, E.; Sabahi, M. An interleaved high step-up converter with coupled inductor and built-in transformer voltage multiplier cell techniques. *IEEE Trans. Ind. Electron.* **2019**, *66*, 1894–1905. [CrossRef]
28. Li, W.; Zhao, Y.; Wu, J.; He, X. Interleaved high step-up converter with winding-cross-coupled inductors and voltage multiplier cells. *IEEE Trans. Power Electron.* **2012**, *27*, 133–143. [CrossRef]
29. He, L.; Liao, Y. An advanced current-autobalance high step-up converter with a multicoupled inductor and voltage multiplier for a renewable power generation system. *IEEE Trans. Power Electron.* **2016**, *31*, 6992–7005.

30. Nouri, T.; Hosseini, S.H.; Babaei, E.; Ebrahimi, J. Interleaved high step-up DC–DC converter based on three-winding high-frequency coupled inductor and voltage multiplier cell. *IET Power Electron.* **2015**, *8*, 175–189. [CrossRef]

31. Venable, D. The *K* factor: A new mathematical tool for stability analysis and synthesis. *Proc. Powercon* **1983**, *10*. Available online: http://citeseerx.ist.psu.edu/viewdoc/download?doi=10.1.1.196.6850&rep=rep1&type=pdf (accessed on 10 May 2020).

32. Cao, L. Type III compensator design for power converters. *Power Electron. Technol.* **2011**, *1*, 20–25.

33. Chen, S.J.; Yang, S.P.; Huang, C.M.; Chen, Y.H. High step-up interleaved converter with three-winding coupled inductors and voltage multiplier cells. In Proceedings of the 20th IEEE International Conference on Industrial Technology (ICIT), Melbourne, VIC, Australia, 13–15 February 2019; pp. 458–463.

Article

A Multi-Input-Port Bidirectional DC/DC Converter for DC Microgrid Energy Storage System Applications

Binxin Zhu [1,2,*], **Hui Hu** [1,2], **Hui Wang** [1,2] **and Yang Li** [3]

1   College of Electrical Engineering and New Energy, China Three Gorges University, Yichang 443000, China; huhuictgu@163.com (H.H.); wanghui@ctgu.edu.cn (H.W.)
2   Hubei Provincial Key Laboratory for Operation and Control of Cascaded Hydropower Station, China Three Gorges University, Yichang 443000, China
3   School of Automation, Wuhan University of Technology, Wuhan 430000, China; yang.li@whut.edu.cn
*   Correspondence: zhubinxin@ctgu.edu.cn; Tel.: +86-1719-765-0066

Received: 23 April 2020; Accepted: 26 May 2020; Published: 1 June 2020

Abstract: A multi-input-port bidirectional DC/DC converter is proposed in this paper for the energy storage systems in DC microgrid. The converter can connect various energy storage batteries to the DC bus at the same time. The proposed converter also has the advantages of low switch voltage stress and high voltage conversion gain. The working principle and performance characteristics of the converter were analyzed in detail, and a 200 W, two-input-port experimental prototype was built. The experimental results are consistent with the theoretical analysis.

Keywords: DC/DC converter; multi-input-port; bidirectional; energy storage

---

1. Introduction

Due to global issues like the greenhouse effect and energy shortage, renewable energy generation has developed rapidly in recent years [1–3]. Renewable energy generation is greatly affected by natural environmental factors, output power of which exhibits intermittence and randomness [4,5]. DC microgrid and energy storage systems, like batteries and supercapacitors, are usually used to smooth the fluctuating and stochastic output power of the renewable energy generation system [6,7]. A DC/DC converter with the capability of bidirectional energy conversion is the key device to connect batteries and the DC bus of the DC microgrid.

In recent years, many studies have been conducted on bidirectional DC/DC converters [8,9]. Many battery cells were connected in series to achieve high voltage [10]; however, a charge equalization circuit needs to be introduced to solve the problem of unbalanced battery charging [11]. On the contrary, many batteries can also be connected in parallel to achieve high reliability [12], but the output voltage of these batteries is low, and a high voltage gain converter is required in such an application [13,14]. Coupled inductors, switch capacitors, or voltage multiple cells can be used to improve the voltage conversion ratio [15–19]; however, most of the above converters are single input and single output, which means a large number of converters have to be used to connect each battery energy storage unit to the DC bus respectively [20,21], as Figure 1a shows.

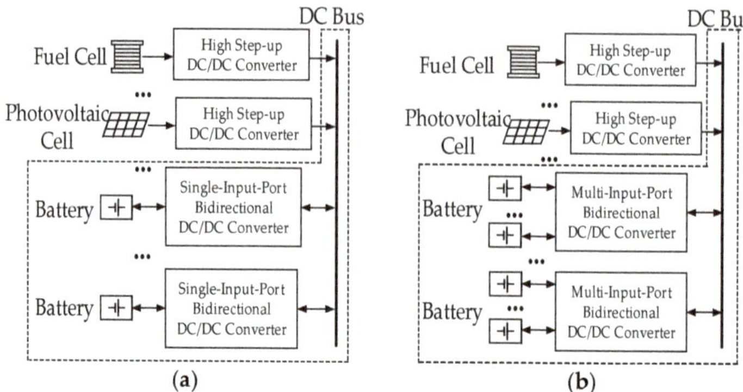

Figure 1. A DC microgrid with various battery energy storage systems. (**a**) traditional converters; (**b**) proposed converter.

In [22–24], some multi-input-port bidirectional converters have been presented; however, these converters have some common disadvantages, such as a large number of devices, large size, and high cost. A multi-input-port bidirectional DC/DC converter is proposed in this paper, many battery energy storage units can be connected to the DC bus by this converter together, as Figure 1b shows. Both in charging and discharging mode, the power flow to every battery can be controlled easily. Apparently, the cost of the whole system can be reduced.

The paper is organized as follows. The working principle, performance analysis, and extension of the proposed converter are described in Sections 2–4, respectively. In Section 5, the efficacy of the proposed converter is verified experimentally using a 200 W prototype.

2. Operation Principle of the Proposed Multi-Input-Port Bidirectional DC/DC Converter

The operation principle of the proposed converter will be presented in this section based on a topology with two input ports shown in Figure 2. To simplify the analysis, the following assumptions are made:

1. The currents i_{L1} and i_{L2} of the inductors L_1 and L_2 are both continuous.
2. All devices are ideal, regardless of the influence of parasitic parameters.
3. The switches S_1 and S_2 are regulated by an interleaved control strategy with the duty cycle greater than 0.5. While the switches Q_1 and Q_2 are controlled by an interleaved control strategy with the duty cycle less than 0.5. The operation principle of the converter can be analyzed based on the discharging or charging modes.

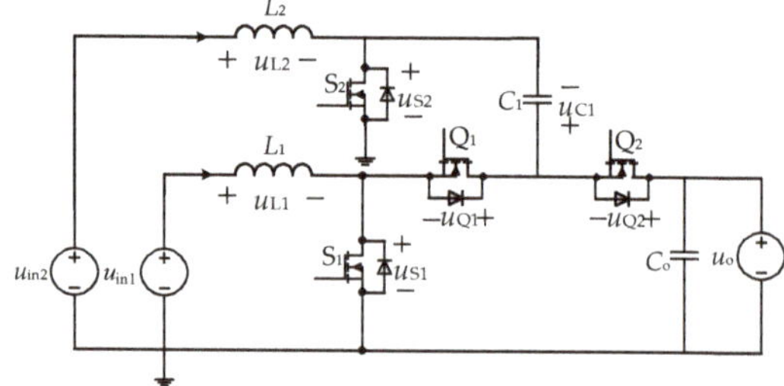

Figure 2. A multi-input-port bidirectional DC/DC converter for energy storage systems in a DC microgrid.

2.1. Discharging Mode (Boost)

In this mode, S_1 and S_2 are interleaved with 180° phase shift to turn on, and Q_1, Q_2 are turned off. During a switching period T_s, there are three Sub-modes. The main waveforms of the converter working in steady state are shown in Figure 3, and the equivalent circuit of each Sub-mode is shown in Figure 4. The control signals of S_1 and S_2 are denoted by u_{gs1} and u_{gs2}, respectively.

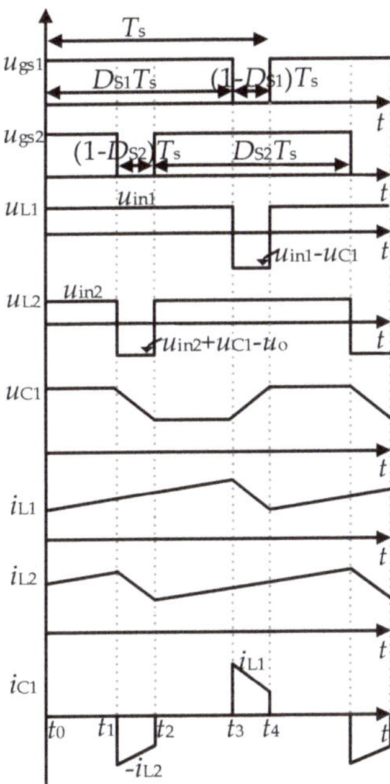

Figure 3. The main waveforms in one switching period T_s.

Sub-mode 1 [t_0–t_1, t_2–t_3]: as Figure 4a shows, S_1 and S_2 are on. The voltages of the inductors L_1 and L_2 are equal to u_{in1} and u_{in2}, respectively. The inductor currents increase linearly at the rates of u_{in1}/L_1 and u_{in2}/L_2, respectively. The current through the capacitor C_1 is zero, while the capacitor voltage is unchanged.

Sub-mode 2 [t_1–t_2]: as Figure 4b shows, S_1 is on, and S_2 is off. Same as Sub-mode 1, the voltage of the inductor L_1 is still u_{in1}, and the current through it increases linearly at the rate of u_{in1}/L_1. However, the current through the inductor L_2 decreases at the rate of $(u_{in2} + u_{C1} - u_o)/L_2$. The capacitor C_1 is being discharged. The voltage of C_1 decreases linearly, and the current of C_1 is equal to i_{L2}.

Sub-mode 3 [t_3–t_4]: as Figure 4c shows, S_1 is off, and S_2 is on. The current through the inductor L_1 decreases at the rate of $(u_{in1} - u_{C1})/L_1$. The voltage of the inductor L_2 is u_{in2}, and the current of L_2 increases at the rate of u_{in2}/L_2. The capacitor C_1 is being charged. The current of the capacitor C_1 is equal to i_{L1}, and the voltage of C_1 increases linearly.

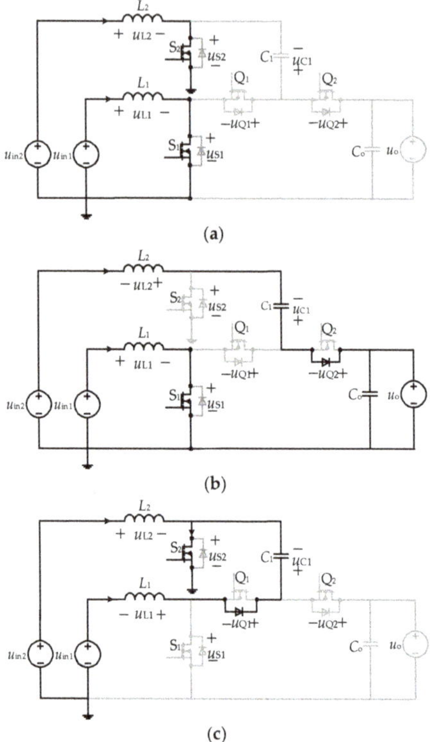

Figure 4. The equivalent circuits in the discharging mode for (**a**) Sub-mode 1; (**b**) Sub-mode 2; (**c**) Sub-mode 3.

2.2. Charging Mode (Buck)

In this mode, Q_1 and Q_2 are interleaved with 180° phase shift to turn on, and S_1, S_2 are off. During a switching period T_s, there are three Sub-modes. The main waveforms of the converter working in steady state are shown in Figure 5, and the equivalent circuit of each Sub-mode is shown in Figure 6. The control signals of Q_1 and Q_2 are denoted by u_{gQ1} and u_{gQ2}, respectively.

Sub-mode 1 [t_0–t_1]: as Figure 6a shows, Q_1 is on, and Q_2 is off. The current through the inductor L_1 increases at the rate of $(u_{C1} - u_{in1})/L_1$. The voltage of the inductor L_2 is u_{in2}, and the current of L_2 decreases at the rate of u_{in2}/L_2. The capacitor C_1 is being discharged. The voltage of C_1 decreases linearly and the current of C_1 is equal to i_{L1}.

Sub-mode 2 [t_1–t_2, t_3–t_4]: as Figure 6b shows, Q_1 and Q_2 are off. The voltages of the inductors L_1 and L_2 are u_{in1} and u_{in2}, respectively. The inductor currents decrease linearly at the rates of u_{in1}/L_1 and u_{in2}/L_2, respectively. The current through the capacitor C_1 is zero, while the capacitor voltage is unchanged.

Sub-mode 3 [t_2–t_3]: as Figure 6c shows, Q_1 is off, and Q_2 is on. The current of the inductor L_1 decreases at the rate of u_{in1}/L_1. However, the current through the inductor L_2 increases at the rate of $(u_o - u_{in2} - u_{C1})/L_2$. The capacitor C_1 is being charged. The current of the capacitor C_1 is equal to i_{L2}, and the voltage of C_1 increases linearly.

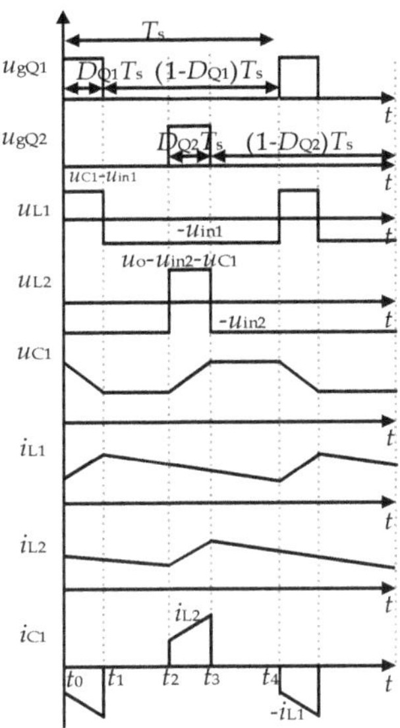

Figure 5. The main waveforms in one switching period T_s.

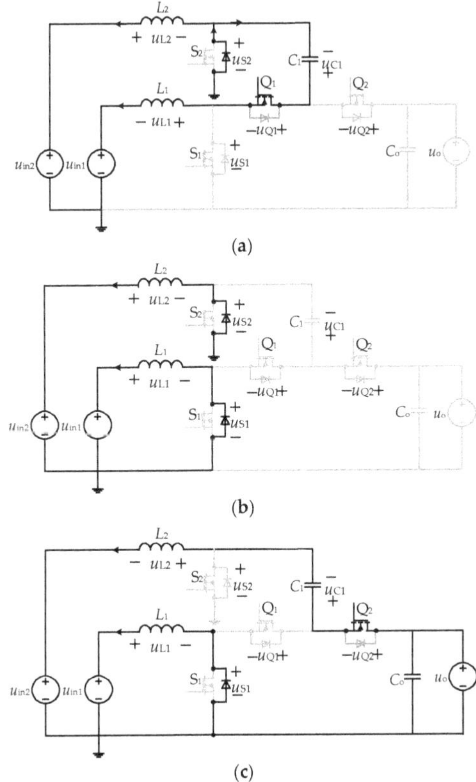

Figure 6. The equivalent circuits in the charging mode for (**a**) Sub-mode 1; (**b**) Sub-mode 2; (**c**) Sub-mode 3.

3. Performance Analysis

3.1. Voltage Conversion Ratio

Discharging Mode (Boost): According to the analysis of the above working principle, the operating characteristics of the proposed converter can be derived from the three Sub-modes in one switching cycle T_s, based on the voltage-second balance of the inductors L_1 and L_2.

$$D_{S1}u_{in1} + (1 - D_{S1})(u_{in1} - u_{C1}) = 0 \tag{1}$$

$$D_{S2}u_{in2} + (1 - D_{S2})(u_{in2} + u_{C1} - u_o) = 0 \tag{2}$$

From Equations (1) and (2), Equations (3) and (4) can be derived:

$$u_{C1} = \frac{u_{in1}}{1 - D_{S1}} \tag{3}$$

$$u_o = \frac{u_{in1}}{1 - D_{S1}} + \frac{u_{in2}}{1 - D_{S2}} \tag{4}$$

According to Equation (4), it can be clearly seen that the voltage conversion ratio of the proposed converter is twice that of the traditional boost converter.

When the input voltages u_{in1}, u_{in2}, and the duty cycle D_{S1}, D_{S2} are the same, respectively, the voltage conversion ratio of the proposed converter can be derived:

$$M_{Boost} = \frac{u_o}{u_{in}} = \frac{2}{1 - D_{Boost}} \tag{5}$$

Charging Mode (Buck): According to the analysis of the above working principle, the operating characteristics of the proposed converter can be derived from the Sub-three modes in one switching cycle T_s, based on the voltage-second balance of the inductors L_1 and L_2.

$$D_{Q1}(u_{C1} - u_{in1}) + (1 - D_{Q1})(-u_{in1}) = 0 \tag{6}$$

$$D_{Q2}(u_o - u_{in2} - u_{C1}) + (1 - D_{Q2})(-u_{in2}) = 0 \tag{7}$$

From Equations (6) and (7), Equations (8) and (9) can be derived:

$$u_{C1} = \frac{u_{in1}}{D_{Q1}} \tag{8}$$

$$u_o = \frac{u_{in1}}{D_{Q1}} + \frac{u_{in2}}{D_{Q2}} \tag{9}$$

When the output voltages u_{in1}, u_{in2}, and the duty cycle D_{Q1}, D_{Q2} are the same, respectively, the voltage conversion ratio of the proposed converter can be derived:

$$M_{Buck} = \frac{u_{in}}{u_o} = \frac{D_{Buck}}{2} \tag{10}$$

According to Equation (10), it can be seen that the voltage conversion ratio of the proposed converter is half of that of the traditional buck converter.

3.2. Relationship between the Currents of the Two Inductors

Discharging Mode (Boost): During a switching cycle T_s, in Sub-mode 3, the capacitor C_1 is charged for $(1 - D_{S1})T_s$ and the current of C_1 is equal to i_{L1}. In Sub-mode 2, the capacitor C_1 is discharged for

$(1 - D_{S2})T_s$, and the current of C_1 is equal to i_{L2}. In Sub-mode 1, the current of the capacitor C_1 is zero. Due to the ampere-second balance of the capacitor C_1, the following can be derived:

$$I_{L1}(1 - D_{S1})T_s = I_{L2}(1 - D_{S2})T_s \tag{11}$$

$$I_{L1}(1 - D_{S1}) = I_{L2}(1 - D_{S2}) \tag{12}$$

When the duty cycles D_{S1} and D_{S2} are equal, the two input currents are also equal. Thus, automatic current sharing is realized. The power of the two ports can be adjusted through controlling D_{S1} and D_{S2}, respectively.

Charging Mode (Buck): During a switching cycle T_s, in Sub-mode 1, the capacitor C_1 is discharged for $D_{Q1}T_s$, and the current of C_1 is equal to i_{L1}. In Sub-mode 3, the capacitor C_1 is charged for $D_{Q2}T_s$, and the current of C_1 is equal to i_{L2}. In Sub-mode 2, the current of the capacitor C_1 is zero. Due to the ampere-second balance of the capacitor C_1, the following can be derived:

$$I_{L1}D_{Q1}T_s = I_{L2}D_{Q2}T_s \tag{13}$$

$$I_{L1}D_{Q1} = I_{L2}D_{Q2} \tag{14}$$

When the duty cycle D_{Q1} and D_{Q2} are equal, the two input currents are also equal. Thus, automatic current sharing is realized. The power of the two ports can be adjusted through controlling D_{Q1} and D_{Q2}, respectively.

3.3. Voltage Stress of Switch

The voltage stresses of S_1, S_2, Q_1, and Q_2 can be derived as follows:

1. Discharging Mode (Boost):

$$u_{S1} = u_{C1} \tag{15}$$

$$u_{S2} = u_{Q2} = u_o - u_{C1} \tag{16}$$

$$u_{Q1} = u_o \tag{17}$$

2. Charging Mode (Buck):

$$u_{S1} = u_{C1} \tag{18}$$

$$u_{S2} = u_{Q2} = u_o - u_{C1} \tag{19}$$

$$u_{Q1} = u_o \tag{20}$$

3.4. Current Stress of Switch

Discharging Mode (Boost): To begin with the time of S_1 turning on, in the following cycle T_s, the inductor currents i_{L1}, i_{L2} can be represented as

$$i_{L1} = \begin{cases} I_{L1} - \dfrac{u_{in1}D_{S1}T_s}{2L_1} + \dfrac{u_{in1}}{L_1}t, & 0 < t \le D_{S1}T_s \\[2mm] I_{L1} + \dfrac{u_{in1}D_{S1}T_s}{2L_1} - \dfrac{u_{C1}-u_{in1}}{L_1}t, & D_{S1}T_s < t \le T_s \end{cases} \tag{21}$$

$$i_{L2} = \begin{cases} I_{L2} + \dfrac{1-D_{S2}}{2L_2}u_{in2}T_s + \dfrac{u_{in2}}{L_2}t, & 0 < t \le (D_{S2} - \frac{1}{2})T_s \\[2mm] I_{L2} + \dfrac{u_{in2}D_{S2}T_s}{2L_2} - \dfrac{u_o-u_{C1}-u_{in2}}{L_2}[t - (D_{S2} - \frac{1}{2})T_s], & (D_{S2} - \frac{1}{2})T_s < t \le \frac{T_s}{2} \\[2mm] I_{L2} - \dfrac{u_{in2}D_{S2}T_s}{2L_2} + \dfrac{u_{in2}}{L_2}(t - \frac{T_s}{2}), & \frac{T_s}{2} < t \le T_s \end{cases} \tag{22}$$

According to the three Sub-modes of the circuit, in a switching cycle, the currents through S_1 and S_2 at every stage can be derived as follows:

$$i_{S1} = \begin{cases} i_{L1}, & 0 < t \le D_{S1}T_s \\ 0, & D_{S1}T_s < t \le T_s \end{cases} \tag{23}$$

$$i_{S2} = \begin{cases} i_{L2}, & 0 < t \le (D_{S2} - \tfrac{1}{2})T_s \\ 0, & (D_{S2} - \tfrac{1}{2})T_s < t \le \tfrac{T_s}{2} \\ i_{L2}, & \tfrac{T_s}{2} < t \le D_{S2}T_s \\ i_{L1} + i_{L2}, & D_{S2}T_s < t \le T_s \end{cases} \tag{24}$$

From Equations (21)–(24), the currents through S_1, S_2, Q_1, and Q_2 can be derived as follows:

$$i_{S1} = i_{Q1} = I_{L1} + \frac{u_{in1}D_{S1}T_s}{2L_1} \tag{25}$$

$$i_{S2} = I_{L1} + \frac{u_{in1}D_{S1}T_s}{2L_1} + I_{L2} + \frac{D_{S2} - 1}{2L_2}u_{in2}T_s \tag{26}$$

$$i_{Q2} = I_{L2} + \frac{u_{in2}D_{S2}T_s}{2L_2} \tag{27}$$

Charging Mode (Buck): To begin with the time of Q_1 turning on, in the following cycle T_s, the inductor currents i_{L1} and i_{L2} can be denoted by

$$i_{L1} = \begin{cases} \frac{u_{C1} - u_{in1}}{L_1}t, & 0 < t \le D_{Q1}T_s \\ I_{L1} + \frac{(u_{C1} - u_{in1})D_{Q1}T_s}{2L_1} - \frac{u_{in1}}{L_1}(t - D_{Q1}T_s), & D_{Q1}T_s < t \le T_s \end{cases} \tag{28}$$

$$i_{L2} = \begin{cases} I_{L2} + \frac{(u_o - u_{C1} - u_{in2})D_{Q2}T_s}{2L_2} - \frac{u_{in2}}{L_2}(\tfrac{1}{2} - D_{Q2})T_s - \frac{u_{in2}}{L_2}t, & 0 < t \le \tfrac{1}{2}T_s \\ I_{L2} - \frac{(u_o - u_{C1} - u_{in2})D_{Q2}T_s}{2L_2} + \frac{(u_o - u_{C1} - u_{in2})}{L_2}(t - \tfrac{1}{2}T_s), & \tfrac{1}{2}T_s < t \le (\tfrac{1}{2} + D_{Q2})T_s \\ I_{L2} + \frac{(u_o - u_{C1} - u_{in2})D_{Q2}T_s}{2L_2} - \frac{u_{in2}}{L_2}[t - (\tfrac{1}{2} + D_{Q2})T_s], & (\tfrac{1}{2} + D_{Q2})T_s < t \le T_s \end{cases} \tag{29}$$

According to the three Sub-modes of the circuit, in a switching cycle, the currents through Q_1 and Q_2 at each stage can be derived as follows:

$$i_{Q1} = \begin{cases} i_{L1}, & 0 < t \le D_{Q1}T_s \\ 0, & D_{Q1}T_s < t \le T_s \end{cases} \tag{30}$$

$$i_{Q2} = \begin{cases} 0, & 0 < t \le \tfrac{1}{2}T_s \\ i_{L2}, & \tfrac{1}{2}T_s < t \le (\tfrac{1}{2} + D_{Q2})T_s \\ 0, & (\tfrac{1}{2} + D_{Q2})T_s < t \le T_s \end{cases} \tag{31}$$

From Equations (28)–(31), the currents through S_1, S_2, Q_1, and Q_2 can be derived as follows:

$$i_{S1} = i_{Q1} = I_{L1} + \frac{u_{in1}\left(1 - D_{Q1}\right)T_s}{2L_1} \tag{32}$$

$$i_{Q2} = I_{L2} + \frac{u_{in2}\left(1 - D_{Q2}\right)T_s}{2L_2} \tag{33}$$

$$i_{S2} = I_{L1} + \frac{u_{in1}\left(1 - D_{Q1}\right)T_s}{2L_1} + I_{L2} - \frac{u_{in2}\left(1 - D_{Q2}\right)T_s}{2L_2} + \frac{u_{in2}}{L_2}\left(\tfrac{1}{2} - D_{Q1}\right)T_s \tag{34}$$

3.5. Power Flow

Discharging Mode (Boost): The inductor current i_{L1} increases as the duty cycle D_{S1} increases, and the inductor current i_{L2} decreases as the duty cycle D_{S2} decreases. Since the two input voltages u_{in1}, u_{in2} are equal, the ratio of the power of the two ports is equal to the ratio of the two inductor currents. Therefore, when $D_{S1} < D_{S2}$, $i_{L1} < i_{L2}$, $i_{L1}/i_{L2} < 1$; when $D_{S1} = D_{S2}$, $i_{L1} = i_{L2}$, $i_{L1}/i_{L2} = 1$; when $D_{S1} > D_{S2}$, $i_{L1} > i_{L2}$, $i_{L1}/i_{L2} > 1$. Making D_{S1}: 0.5–0.8 as the *x*-axis, D_{S2}: 0.8–0.5 as the *y*-axis, and i_{L1}/i_{L2} as the *z*-axis, the following three-dimensional figure can be obtained as Figure 7 shows.

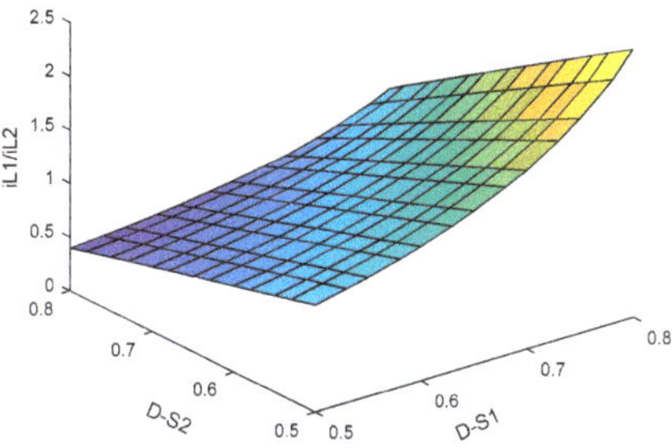

Figure 7. i_{L1}/i_{L2} 3D graph (Discharging Mode (Boost)).

Charging Mode (Buck): The inductor current i_{L1} increases as the duty cycle D_{Q1} decreases, and the inductor current i_{L2} decreases as the duty cycle D_{Q2} increases. Since the two output voltages u_{in1}, u_{in2} are equal, the ratio of the power of the two ports is equal to the ratio of the two inductor currents. Therefore, when $D_{Q1} < D_{Q2}$, $i_{L1} > i_{L2}$, $i_{L1}/i_{L2} > 1$; when $D_{Q1} = D_{Q2}$, $i_{L1} = i_{L2}$, $i_{L1}/i_{L2} = 1$; when $D_{Q1} > D_{Q2}$, $i_{L1} < i_{L2}$, $i_{L1}/i_{L2} < 1$. Making D_{Q1}: 0.5–0.2 as the *x*-axis, D_{Q2}: 0.2–0.5 as the *y*-axis, and i_{L1}/i_{L2} as the *z*-axis, the following three-dimensional figure can be obtained as Figure 8 shows.

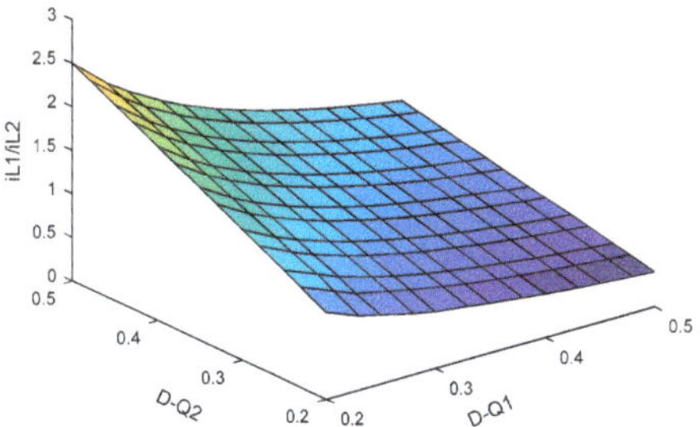

Figure 8. i_{L1}/i_{L2} 3D graph (Charging Mode (Buck)).

3.6. Comparison of the Proposed Converter with Other Converters

Some quantitative comparisons between some existing multi-input-port topologies, and the proposed converter are given in Table 1. As can be seen, compared to [22–24], the number of devices of the proposed converter is less, which means fewer losses and a lower cost.

Table 1. Comparison of the converters.

	[22]	[23]	[24]	Proposed
No. of ports	3	4	3	3
No. of switches	12	4	6	4
No. of diodes	0	4	0	0
No. of inductors	6	4	2	2
No. of capacitors	3	5	2	2

4. Extension of the Topology

4.1. Topology of the N-Input-Port Bidirectional DC/DC Converter

Based on the topology of the two-input-port bidirectional DC/DC converter shown in Figure 2, the n-input-port bidirectional DC/DC converter topology can be derived as Figure 9 shows. To simplify, assumptions are made as follows:

1. Currents of the inductors i_{L1}, i_{L2}, ... , and i_{L2} are all continuous.
2. All devices are ideal, regardless of the influence of parasitic parameters.
3. Discharging Mode (Boost): during a switching period T_s, S_1, S_2, ..., and S_n interleaved with 360°/n phase shift are turned on with the duty cycle greater than $(1 - 1/n)$, and Q_1, Q_2, ..., and Q_n are turned off. Charging Mode (Buck): during a switching period T_s, Q_1, Q_2, ..., and Q_n interleaved with 360°/n phase shift are turned on with the duty cycle less than $1/n$, and S_1, S_2, ..., and S_n are turned off.

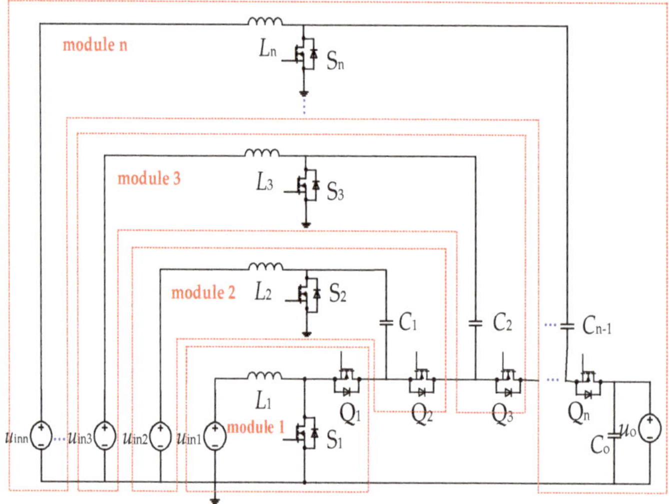

Figure 9. An n-input-port bidirectional DC/DC converter for DC microgrid energy storage system.

4.2. Voltage Conversion Ratio

Discharging Mode (Boost): Due to the voltage-second balance of the inductors L_1, L_2, ... , and L_n, it can be derived:

$$D_{S(i-1)}u_{\mathrm{in}(i-1)} = (1 - D_{S(i-1)})(u_{C(i-1)} - u_{C(i-2)} - u_{\mathrm{in}(i-1)}) \tag{35}$$

$$D_{Si}u_{\mathrm{in}i} = (1 - D_{Si})(u_o - u_{C(i-1)} - u_{\mathrm{in}i}) \tag{36}$$

$$u_{Ci} = \sum_{p=1}^{i} \frac{u_{\mathrm{in}p}}{1 - D_{Sp}} \quad (1 \le i \le n-1) \tag{37}$$

$$u_o = \sum_{i=1}^{n} \frac{u_{\mathrm{in}i}}{1 - D_{Si}} \tag{38}$$

When $D_{S1} = D_{S2} = \ldots = D_{sn} = D_{\mathrm{Boost}}$, the ratio of the output voltage u_o, and each input voltage $u_{\mathrm{in}i}$ is the voltage gain M_i of each input port.

$$M_i = \frac{u_o}{u_{\mathrm{in}i}} \tag{39}$$

$$\sum_{i=1}^{n} \frac{u_{\text{in}i}}{u_{\text{o}}} = 1 - D_{\text{Boost}} \tag{40}$$

$$\frac{1}{M_1} + \frac{1}{M_2} + \cdots + \frac{1}{M_n} = 1 - D_{\text{Boost}} \tag{41}$$

When $u_{\text{in}1} = u_{\text{in}2} = \ldots = u_{\text{in}n}$,

$$M_1 = M_2 = \cdots = M_n = \frac{n}{1 - D_{\text{Boost}}} \tag{42}$$

Charging Mode (Buck): Due to the voltage-second balance of the inductors $L_1, L_2, \ldots$, and L_n, it can be derived:

$$(1 - D_{Q(i-1)})u_{\text{in}(i-1)} = D_{Q(i-1)}\left(u_{C(i-1)} - u_{C(i-2)} - u_{\text{in}(i-1)}\right) \tag{43}$$

$$(1 - D_{Qi})u_{\text{in}i} = D_{Qi}\left(u_{\text{o}} - u_{C(i-1)} - u_{\text{in}i}\right) \tag{44}$$

$$u_{Ci} = \sum_{p=1}^{i} \frac{u_{\text{in}p}}{D_{Qp}}(1 \leq i \leq n - 1) \tag{45}$$

$$u_{\text{o}} = \sum_{i=1}^{n} \frac{u_{\text{in}i}}{D_{Qi}} \tag{46}$$

When $D_{Q1} = D_{Q2} = \ldots = D_{Qn} = D_{\text{Buck}}$, the ratio of each output voltage $u_{\text{in}i}$ and the input voltage u_{o} is the voltage gain M_i of each output port.

$$M_i = \frac{u_{\text{in}i}}{u_{\text{o}}} \tag{47}$$

$$\sum_{i=1}^{n} \frac{u_{\text{in}i}}{u_{\text{o}}} = D_{\text{Buck}} \tag{48}$$

$$M_1 + M_2 + \cdots + M_n = D_{\text{Buck}} \tag{49}$$

When $u_{\text{in}1} = u_{\text{in}2} = \ldots = u_{\text{in}n}$,

$$M_1 = M_2 = \cdots = M_n = \frac{D_{\text{Buck}}}{n} \tag{50}$$

4.3. Relationship between the Currents of the Inductors

It is assumed that the average values of the inductor currents $i_{L1}, i_{L2}, \ldots$, and i_{Ln} are $I_{L1}, I_{L2}, \ldots$, and I_{Ln}, respectively.

Discharging Mode (Boost): Due to the ampere-second balance of the capacitors $C_1, C_2, \ldots$, and C_n, it can be derived as follows:

$$I_{L1}(1 - D_{S1}) = I_{L2}(1 - D_{S2}) = \cdots = I_{Ln}(1 - D_{Sn}) \tag{51}$$

When $D_{S1} = D_{S2} = \ldots = D_{Sn}$, $I_{L1} = I_{L2} = \ldots = I_{Ln}$. Thus, automatic current sharing is realized. The power of all the ports can be adjusted through controlling $D_{S1}, D_{S2}, \ldots$, and D_{Sn}, respectively.

Charging Mode (Buck): Due to the ampere-second balance of the capacitors $C_1, C_2, \ldots$, and C_n, it can be derived:

$$I_{L1}D_{Q1} = I_{L2}D_{Q2} = \cdots = I_{Ln}D_{Qn} \tag{52}$$

When $D_{Q1} = D_{Q2} = \ldots = D_{Qn}$, $I_{L1} = I_{L2} = \ldots = I_{Ln}$. Thus, automatic current sharing is realized. The power of all the ports can be adjusted through controlling $D_{Q1}, D_{Q2}, \ldots$, and D_{Qn}, respectively.

In practical applications, the efficiency of the converter will drop along with the increase in the number of input ports.

5. Experimental Results

To verify the analysis presented in the previous sections, experiments were conducted based on a 200 W two-input-port prototype developed from the proposed converter. The specification of the prototype is given in Table 2, and the experimental results are presented and discussed as follows.

Table 2. Specification of the prototype.

Parameters	Values
Voltage (u_{in1}, u_{in2})	24 V
Voltage (u_o)	200 V
Output power (P_o)	200 W
Switching frequency (f_s)	100 kHz
Switch (S_1, S_2, Q_1, Q_2)	C3M0280090D
Capacitors	C_o: 10 uF, C_1: 4 uF
Inductors (L_1, L_2)	400 uH

5.1. Constant Duty Cycle

Discharging Mode (Boost): Figure 10a shows the waveforms of u_{gs1}, u_{gs2}, u_{in1}, and u_{in2}, where the duty cycles are around 0.76. Figure 10b shows the waveforms of u_o, u_{Co}, and u_{C1}. It can be seen that the DC values of u_{C1} and u_o are about 100 V and 200 V, respectively. The voltage conversion gain is around 8.3, which is consistent with that calculated by Equation (5). Figure 10c shows that the voltage stresses of S_1, S_2, and Q_2 are all about 100 V, while the voltage stress of Q_1 is about 200 V. These are consistent with that obtained by Equations (15)–(17). Figure 10d shows that the currents of L_1 and L_2 are both about 4 A. Apparently, the measured results are all consistent with the previous analysis.

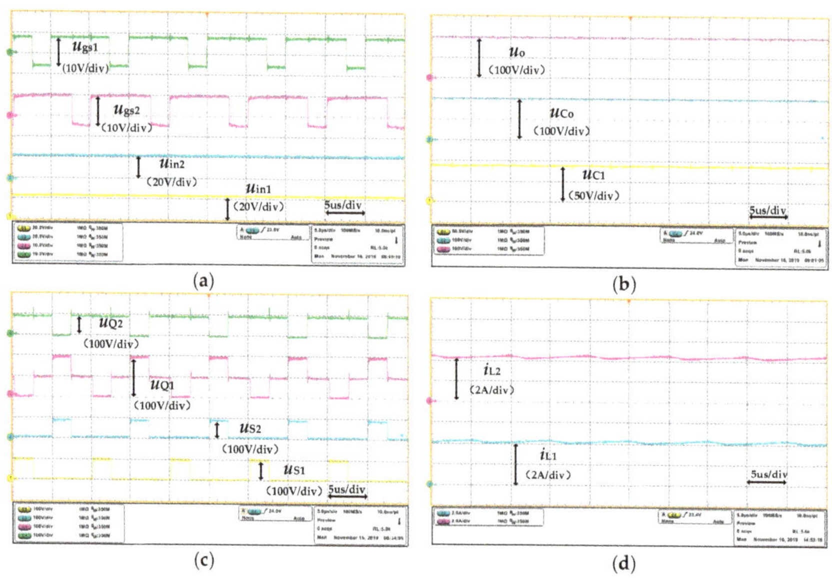

Figure 10. The waveforms of the experimental prototype. (**a**) driving waveforms of the switches and waveforms of the input voltage. (**b**) waveforms of voltages of C_o, C_1, and the waveform of the output voltage. (**c**) waveforms of voltages of the switches. (**d**) waveforms of currents of the inductors.

Charging Mode (Buck): Figure 11a shows the waveforms of u_{gQ1}, u_{gQ2}, u_{in1}, and u_{in2}, the duty cycles are near 0.24. Figure 11b shows the waveforms of u_o, u_{Co}, and u_{C1}; it can be seen that the DC values of u_{C1}, u_o are about 100 V, 200 V, and the conversion gain is approximately 0.12, which is consistent with Equation (10). Figure 11c shows voltage stresses of S_1, S_2, Q_2 are nearly 100 V, and the voltage stress of Q_1 is about 200 V, which are consistent with Equations (18)–(20). Figure 11d shows the waveforms of i_{L1}, i_{L2}. The DC values of i_{L1}, i_{L2} are both about 4 A; evidently, the measured results are all consistent with the theoretical analysis.

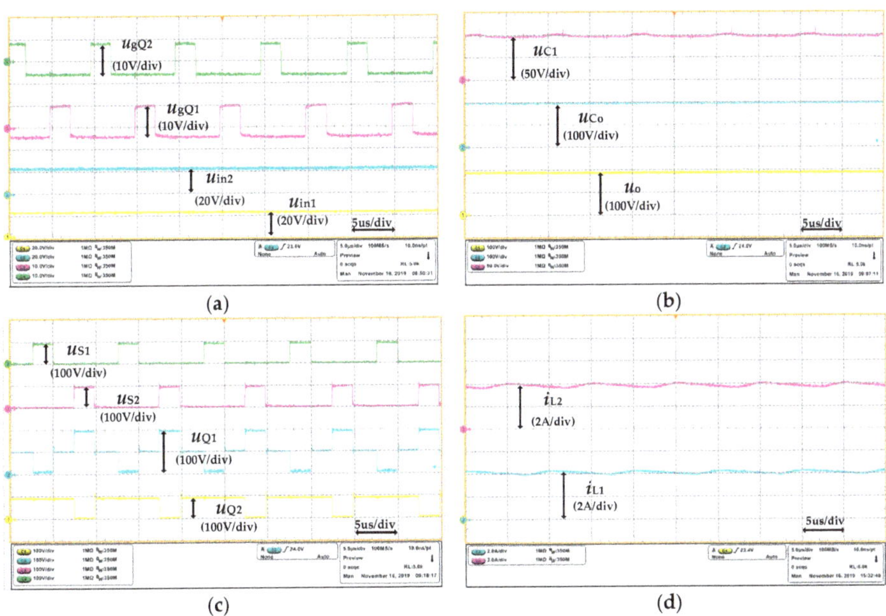

Figure 11. The waveforms of the experimental prototype. (**a**) driving waveforms of the switches and waveforms of the output voltage. (**b**) waveforms of voltages of C_o, C_1, and the waveform of the input voltage. (**c**) waveforms of voltages of the switches. (**d**) waveforms of currents of the inductors.

5.2. Varying Duty Cycle

Discharging Mode (Boost): Figure 12a shows the changes of i_{L1} and i_{L2} when the duty cycle D_{S1} and D_{S2} are adjusted. With the increase of the duty cycle, the inductor current increases, and the power of the branch circuit increases.

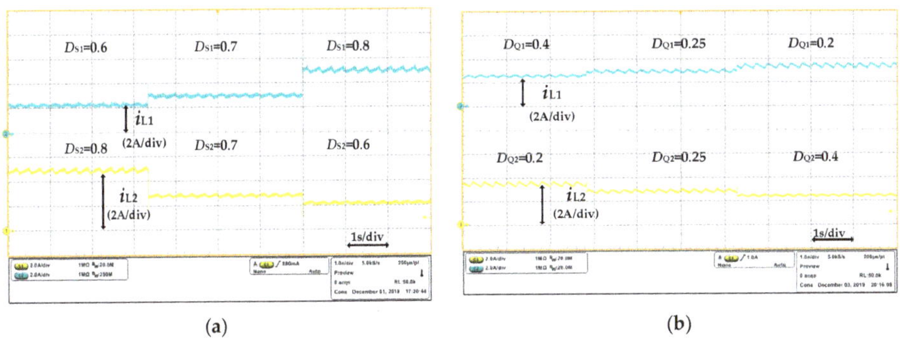

Figure 12. Power flow diagram for (**a**) discharging mode (Boost); (**b**) charging mode (Buck).

Charging Mode (Buck): Figure 12b shows the changes of i_{L1} and i_{L2} when the duty cycle D_{Q1} and D_{Q2} are adjusted. With the increase of the duty cycle, the inductor current decreases, and the power of the branch circuit decreases.

5.3. Converter Efficiency and Conversion Ratio

Based on the experimental results, the converter efficiency and the conversion ratio are analyzed and presented in this sub-section.

Discharging Mode (Boost): Figure 13a shows the curve of efficiency changing with output voltage after changing the duty cycle and the curve of efficiency changing with output power after changing the load. The calculated loss distribution of the experimental prototype is shown in Figure 13b. The main losses are switching losses 9.59 W, anti-parallel diode losses 3.6 W, and inductor losses 2.874 W. As is shown in Figure 13c, the voltage conversion ratio changes with the duty cycle. When the duty cycle is more than 0.7, the difference between the actual gain and the theoretical gain gradually increases as the duty cycle increases.

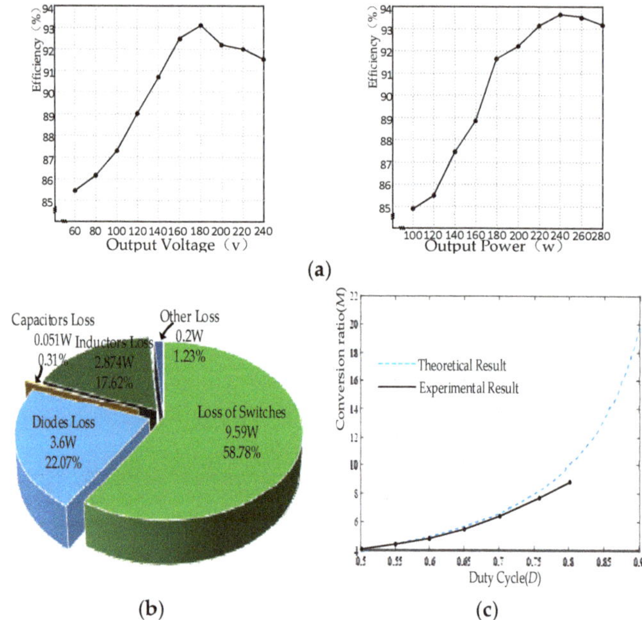

Figure 13. Discharging mode (Boost): (**a**) efficiency curves of the prototype; (**b**) loss distribution of the prototype; (**c**) conversion ratio (*M*) vs. duty cycle (*D*) graph.

Charging Mode (Buck): Figure 14a shows the curve of efficiency changing with output voltage after changing the duty cycle and the curve of efficiency changing with output power after changing the load. The calculated loss distribution of the experimental prototype is shown in Figure 14b. The main losses are anti-parallel diode losses 9 W, switching losses 4.38 W, and inductor losses 2.874 W. As is shown in Figure 14c, the voltage conversion ratio changes with the duty cycle. When the duty cycle is less than 0.3, the difference between the actual gain and the theoretical gain gradually increases as the duty cycle decreases.

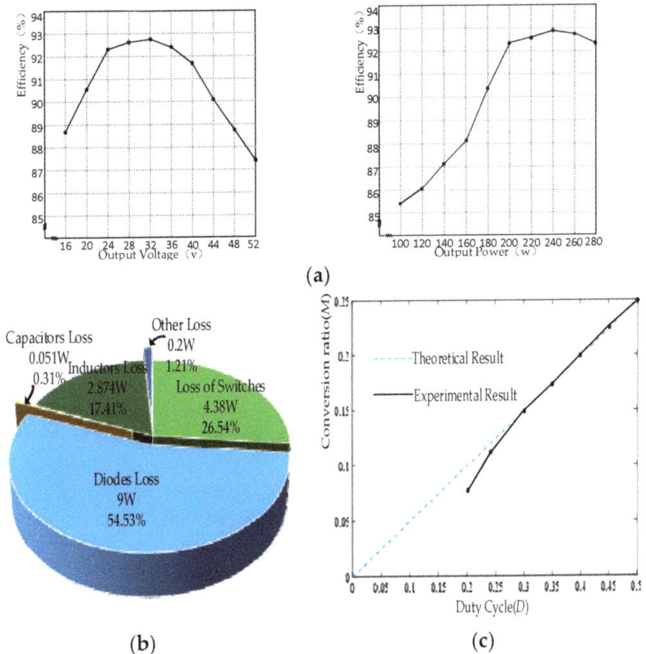

Figure 14. Charging mode (Buck): (**a**) efficiency curves of the prototype; (**b**) loss distribution of the prototype; (**c**) conversion ratio (*M*) vs. duty cycle (*D*) graph.

6. Conclusions

A multi-input-port bidirectional DC/DC converter for DC microgrid energy storage system applications is proposed in this paper. Comprehensive analyses on the working principle and performance of the proposed converter are given. Experimental results are presented, and it is verified that, compared to the traditional buck and boost converter, the proposed bidirectional converter has the following advantages: (1) a wider range of voltage conversion can be achieved and the voltage stresses of the switches are lower; (2) the power flow of each port can be adjusted easily through the controlling of duty cycles; (3) the number of input ports of the proposed converter can be expanded, which makes it more applicable.

Author Contributions: Conceptualization, B.Z.; methodology, B.Z. and H.H.; software, B.Z. and H.H.; writing—original draft preparation, H.H.; writing—review and editing, B.Z. and H.H.; supervision, H.W. and Y.L.; funding acquisition, B.Z. All authors have read and agreed to the published version of the manuscript.

Funding: This research was funded by the National Natural Science Foundation of China, grant number 51707103.

Conflicts of Interest: The authors declare no conflict of interest.

References

1. Wu, Y.-E.; Hsu, K.-C. Novel Three-Port Bidirectional DC/DC Converter with Three-Winding Coupled Inductor for Photovoltaic System. *Energies* **2020**, *13*, 1132. [CrossRef]
2. Zhu, B.; Zeng, Q.; Chen, Y.; Zhao, Y.; Liu, S. A Dual-Input High Step-Up DC/DC Converter with ZVT Auxiliary Circuit. *IEEE Trans. Energy Convers.* **2018**, *34*, 161–169. [CrossRef]
3. Zhu, B.; Ding, F.; Vilathgamuwa, D.M. Coat Circuits for DC–DC Converters to Improve Voltage Conversion Ratio. *IEEE Trans. Power Electron.* **2020**, *35*, 3679–3687. [CrossRef]

4. Wang, C.; Li, X.; Tian, T.; Xu, Z.; Chen, R. Coordinated Control of Passive Transition from Grid-Connected to Islanded Operation for Three/Single-Phase Hybrid Multimicrogrids Considering Speed and Smoothness. *IEEE Trans. Ind. Electron.* **2020**, *67*, 1921–1931. [CrossRef]

5. Li, Y.; Vilathgamuwa, D.M.; Choi, S.S.; Xiong, B.; Tang, J.; Su, Y.; Wang, Y. Design of minimum cost degradation-conscious lithium-ion battery energy storage system to achieve renewable power dispatchability. *Appl. Energy* **2020**, *260*, 114282. [CrossRef]

6. Xu, G.; Xu, L.; Yao, L. Wind turbines output power smoothing using embedded energy storage systems. *J. Mod. Power Syst. Clean Energy* **2013**, *1*, 49–57. [CrossRef]

7. Li, Z.H.; Xiang, X.; Hu, T.H.; Abu-Siada, A.; Li, Z.X.; Xu, Y.C. An improved digital integral algorithm to enhance the measurement accuracy of Rogowski coil-based electronic transformers. *INT J. Elec. Power* **2020**, *118*, 105806. [CrossRef]

8. Lai, C.-M.; Cheng, Y.-H.; Hsieh, M.-H.; Lin, Y.-C. Development of a Bidirectional DC/DC Converter with Dual-Battery Energy Storage for Hybrid Electric Vehicle System. *IEEE Trans. Veh. Technol.* **2018**, *67*, 1036–1052. [CrossRef]

9. Wai, R.-J.; Zhang, Z.-F. Design of High-Efficiency Isolated Bidirectional DC/DC Converter with Single-Input Multiple-Outputs. *IEEE Access* **2019**, *7*, 87543–87560. [CrossRef]

10. Park, H.-S.; Kim, C.-E.; Kim, C.-H.; Moon, G.-W.; Lee, J.-H. A Modularized Charge Equalizer for an HEV Lithium-Ion Battery String. *IEEE Trans. Ind. Electron.* **2009**, *56*, 1464–1476. [CrossRef]

11. Anno, T.; Koizumi, H. Double-Input Bidirectional DC/DC Converter Using Cell-Voltage Equalizer with Flyback Transformer. *IEEE Trans. Power Electron.* **2014**, *30*, 2923–2934. [CrossRef]

12. Ibanez, F.; Echeverria, J.M.; Vadillo, J.; Fontan, L. High-Current Rectifier Topology Applied to a 4-kW Bidirectional DC–DC Converter. *IEEE Trans. Ind. Appl.* **2013**, *50*, 68–77. [CrossRef]

13. Duan, R.-Y.; Lee, J.-D. High-efficiency bidirectional DC-DC converter with coupled inductor. *IET Power Electron.* **2012**, *5*, 115–123. [CrossRef]

14. Zhou, L.-W.; Zhu, B.-X.; Luo, Q.-M. High step-up converter with capacity of multiple input. *IET Power Electron.* **2012**, *5*, 524–531. [CrossRef]

15. Saadat, P.; Abbaszadeh, K. A Single-Switch High Step-Up DC–DC Converter Based on Quadratic Boost. *IEEE Trans. Ind. Electron.* **2016**, *63*, 7733–7742. [CrossRef]

16. Hsieh, Y.-P.; Chen, J.-F.; Yang, L.-S.; Wu, C.-Y.; Liu, W.-S. High-Conversion-Ratio Bidirectional DC–DC Converter with Coupled Inductor. *IEEE Trans. Ind. Electron.* **2013**, *61*, 210–222. [CrossRef]

17. Zhang, Y.; Liu, H.; Li, J.; Sumner, M.; Xia, C. DC–DC Boost Converter with a Wide Input Range and High Voltage Gain for Fuel Cell Vehicles. *IEEE Trans. Power Electron.* **2019**, *34*, 4100–4111. [CrossRef]

18. Wu, G.; Ruan, X.; Ye, Z. Nonisolated High Step-Up DC–DC Converters Adopting Switched-Capacitor Cell. *IEEE Trans. Ind. Electron.* **2014**, *62*, 383–393. [CrossRef]

19. Zhu, B.; Wang, H.; Vilathgamuwa, D.M.; Vilathgamuwa, M. Single-switch high step-up boost converter based on a novel voltage multiplier. *IET Power Electron.* **2019**, *12*, 3732–3738. [CrossRef]

20. Varesi, K.; Hosseini, S.H.; Sabahi, M.; Babaei, E.; Vosoughi, N. Performance and design analysis of an improved non-isolated multiple input buck DC–DC converter. *IET Power Electron.* **2017**, *10*, 1034–1045. [CrossRef]

21. Mangu, B.; Akshatha, S.; Suryanarayana, D.; Fernandes, B.G.; Bhukya, M. Grid-Connected PV-Wind-Battery-Based Multi-Input Transformer-Coupled Bidirectional DC-DC Converter for Household Applications. *IEEE J. Emerg. Sel. Top. Power Electron.* **2016**, *4*, 1086–1095. [CrossRef]

22. Wang, Z.; Li, H. An Integrated Three-Port Bidirectional DC–DC Converter for PV Application on a DC Distribution System. *IEEE Trans. Power Electron.* **2012**, *28*, 4612–4624. [CrossRef]

23. Wu, H.; Xu, P.; Hu, H.; Zhou, Z.; Xing, Y. Multiport Converters Based on Integration of Full-Bridge and Bidirectional DC–DC Topologies for Renewable Generation Systems. *IEEE Trans. Ind. Electron.* **2013**, *61*, 856–869. [CrossRef]

24. Hintz, A.; Prasanna, U.R.; Rajashekara, K. Novel Modular Multiple-Input Bidirectional DC–DC Power Converter (MIPC) for HEV/FCV Application. *IEEE Trans. Ind. Electron.* **2014**, *62*, 3163–3172. [CrossRef]

 *energies*

Article

Evaluation of a Three-Phase Bidirectional Isolated DC-DC Converter with Varying Transformer Configurations Using Phase-Shift Modulation and Burst-Mode Switching

Nuraina Syahira Mohd Sharifuddin [1,*], Nadia M. L. Tan [1,2,*] and Hirofumi Akagi [3]

[1]   Department of Electrical and Electronics Engineering, Universiti Tenaga Nasional, Kajang 43000, Malaysia
[2]   Institute of Power Engineering, Universiti Tenaga Nasional, Kajang 43000, Malaysia
[3]   Innovator and Inventor Development Platform, Tokyo Institute of Technology, Yokohama 226-8502, Japan; akagi@ee.titech.ac.jp
*   Correspondence: SE22704@utn.edu.my (N.S.M.S.); nadia@uniten.edu.my (N.M.L.T.)

Received: 27 April 2020; Accepted: 25 May 2020; Published: 2 June 2020

Abstract: This paper presents the performance of a three-phase bidirectional isolated DC-DC converter (3P-BIDC) in wye-wye (Yy), wye-delta (Yd), delta-wye (Dy), and delta-delta (Dd) transformer configurations, using enhanced switching strategy that combines phase-shift modulation and burst-mode switching. A simulation verification using PSCAD is carried out to study the feasibility and compare the efficiency performance of the 3P-BIDC with each transformer configuration, using intermittent switching, which combines the conventional phase-shift modulation (PSM) and burst-mode switching, in the light load condition. The model is tested with continuous switching that employs the conventional PSM from medium to high loads (greater than 0.3 p.u.) and with intermittent switching at light load (less than 0.3 p.u), in different transformer configurations. In all tests, the DC-link voltages are equal to the transformer turns ratio of 1:1. This paper also presents the power loss estimation in continuous and intermittent switching to verify the modelled losses in the 3P-BIDC in the Yy transformer configuration. The 3P-BIDC is modelled by taking into account the effects that on-state voltage drop in the insulated-gate bipolar transistor (IGBTs) and diodes, snubber capacitors, and three-phase transformer copper winding resistances will have on the conduction and switching losses, and copper losses in the 3P-BIDC. The intermitting switching improves the efficiency of the DC-DC converter with Yy, Yd, Dy, and Dd connections in light-load operation. The 3P-BIDC has the best efficiency performance using Yy and Dd transformer configurations for all power transfer conditions in continuous and intermittent switching. Moreover, the highest efficiency of 99.6% is achieved at the light power transfer of 0.29 p.u. in Yy and Dd transformer configurations. However, the theoretical current stress in the 3P-BIDC with a Dd transformer configuration is high. Operation of the converter with Dy transformer configuration is less favorable due to the efficiency achievements of lower than 95%, despite burst-mode switching being applied.

Keywords: three-phase bidirectional isolated DC-DC converter; burst-mode switching; high-frequency transformer configurations; phase-shift modulation; intermittent switching; three-phase dual-active bridge

---

1. Introduction

The bidirectional isolated DC-DC converter (BIDC), also known as the dual-active bridge (DAB), has become a research interest in recent years [1] for energy storage applications in electric vehicle and renewable energy systems, and solid-state transformers in all-electric-aircraft and ship applications [2–7]. The advantages of a BIDC include bidirectional power flow, small filter components, low device and component stresses, small number of components, and buck-boost operation capability. Many publications have focused on the efficiency improvement of the single-phase BIDC (1P-BIDC) [5,8–13]. However, there is increasing interest in the three-phase BIDC (3P-BIDC) due to the advantages of higher power density, lower switching stresses on the components, minimal backflow power, and higher efficiency compared to the single-phase BIDC [6,7,14–22].

Figure 1 shows the schematic diagram of the 3P-BIDC. It consists of a high-frequency three-phase transformer with a turns ratio of N:1. Bridge 1 is the high-voltage side (HVS) and bridge 2 is the low-voltage side (LVS). The DC-DC converter operates in the buck mode when power is transferred from bridge 1 to 2, and in the boost mode when power is transferred from bridge 2 to 1.

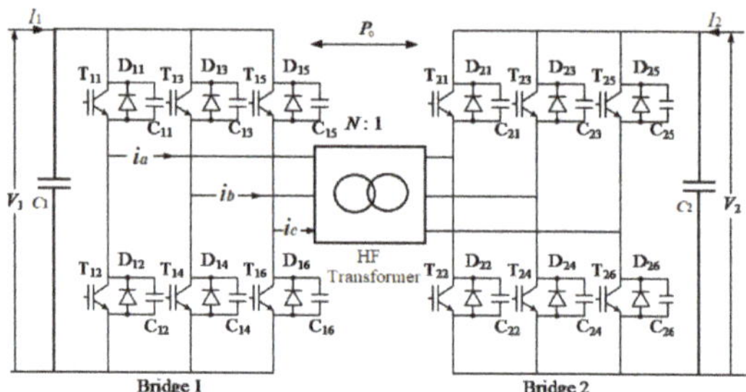

Figure 1. A bidirectional isolated DC-DC converter topology.

Figure 2 illustrates that the HVS and LVS of the transformer can have Yy, Dd, Yd, or Dy configurations. However, a typical configuration for the high-frequency transformer in the BIDC is Yy. An isolated DC-DC converter with Yy transformer connection is shown to have low efficiency levels when not operated in a DC-link voltage ratio of 1:1 [17]. The Dd transformer configuration possesses the same symmetrical characteristics as the Yy transformer configuration and shares the same efficiency characteristics. Symmetrical three-phase windings in Yy and Dd connections have no circulating current flow in the transformer minimizing transformer loss. The 3P-BIDC can be operated in DC-link voltage ratios other than 1:1 with minimized power loss over a wide range of power transfer when the transformer configuration is Yd or Dy [15,16]. Moreover, the 3P-BIDC can operate in zero-voltage switching across the full range of the output current within a certain DC-voltage ratio [15,16].

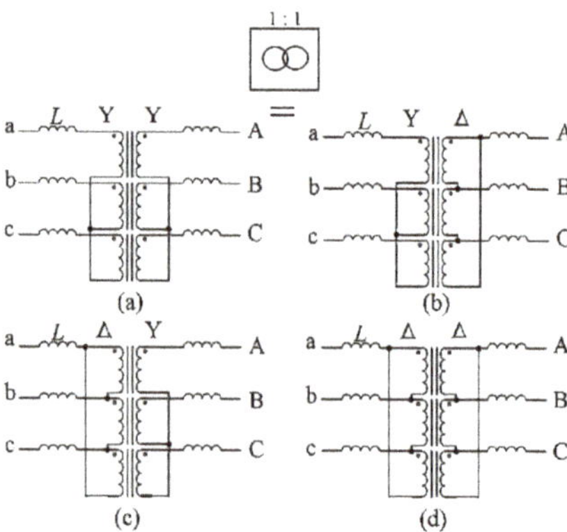

Figure 2. Types of three-phase transformer configurations. (**a**) Yy. (**b**) Yd. (**c**) Dy. (**d**) Dd.

The improvement of efficiency in a 3P-BIDC is also achieved through different switching techniques other than the traditional phase-shift modulation (PSM), such as asymmetrical pulse width modulation cascaded with single-phase shift control [7], triangular and trapezoidal modulation [18], and burst-mode control strategies [6,13,17,20,23–26]. Nevertheless, the high-frequency (HF) transformer configuration in those cases has been Yy connection. There has not been any extensive research based on the performance of a 3P-BIDC that adapts the transformer configurations other than the conventional Yy connection with other switching strategies. The authors of [6] presented experimental results of the 3P-BIDC using burst-mode switching strategy in medium-load operation. However, burst-mode switching did not improve the efficiency of the 3P-BIDC in medium-load operation. The authors further analyzed the performance of the 3P-BIDC in different transformer configurations with the conventional PSM technique [16]. There has been a lack of analysis of the 3P-BIDC in other transformer configurations even though the burst-mode switching has been proposed in many studies [17,20,23–28] and is suitable for light-load efficiency optimization in Yy configuration [27]. Since the potential of further improving the efficiency of the 3P-BIDC is significant, there is a need to investigate the effects of different HF transformer configurations in switching techniques other than PSM techniques, such as the burst-mode switching technique in light-load conditions.

Burst-mode switching technique enables intermittent power transfer during a light-load operation in single-phase and three-phase bidirectional isolated DC-DC converters [17,20]. The burst-mode strategy is a method used to improve the light-load efficiency of power converters by minimizing the switching losses. The method also significantly improves light-load efficiency of other types of DC-DC converters [23–28]. In burst-mode strategy, the transistors are turned ON and OFF cyclically at a certain fixed frequency to produce a burst of pulses that can be transferred to the output.

This paper presents the feasibility of operation and compares the efficiency performance of a 3P-BIDC with different HF transformer configurations, namely, Yy, Yd, Dy, and Dd, when it is operated in continuous and intermittent switching. The 3P-BIDC is modelled in PSCAD by taking into account the effects that on-state voltage drop in the IGBTs and diodes, snubber capacitors, and three-phase transformer copper winding resistances will have on the conduction and switching losses, and copper losses in the 3P-BIDC. Theoretical current stress and loss analyses of the 3P-BIDC in Yy and Dd transformer configuration are also presented. The theoretical loss analysis is coherent with the loss measured in the simulated model.

2. 3P-BIDC with Different Transformer Configurations and Different Power Transfer

The operating principles of the 3P-BIDC based on PSM is explained mode by mode in [17,21]. The power transfer equations used in this section are based on [1,15,16]. In this paper, the 3P-BIDC in Yy, Yd, Dy, and Dd connection are designed to operate in a transformer turns ratio of 1:1. If the DC-link voltage ratio deviates from the transformer turns ratio, the DC-DC converter will not perform well. This type of analysis is presented in [17]. For the sake of simplicity, this paper only discusses the comparison of the 3P-BIDC in different transformer configurations with the DC-link voltage ratios equal to the transformer turns ratio.

2.1. Wye-Wye (Yy) Connection

Figure 2a shows the Yy connection at the HVS and LVS of the HF transformer, which is a typical transformer configuration for the 3P-BIDC in Figure 1. The power transfer equation for the 3P-BIDC in Yy connection for the phase-shift angle range of $0 \leq \delta \leq \frac{\pi}{3}$ is,

$$P_o = P_{Yy} = \frac{V_1 N V_2}{2\pi f_s L_{Yy}} \delta \left(\frac{2}{3} - \frac{\delta}{2\pi} \right), \tag{1}$$

where V_1 and V_2 are the HVS and LVS DC-link voltages, respectively, f_s is the continuous switching frequency, δ is the phase-shift angle between the ac phase voltages of bridges 1 and 2 in radians, N is the transformer turns ratio, and L_{Yy} is the per phase leakage inductance of the transformer in the Yy connection.

2.2. Wye-Delta (Yd) and Delta-Wye (Dy) Connection

Figure 2b presents the three-phase transformer in Yd connection. The HVS of the transformer is connected in wye (Y) and the LVS of the transformer is connected in delta (d). The power transfer equation for phase-shift angles in the range of $0 \leq \delta \leq \frac{\pi}{3}$ is,

$$P_o = P_{Yd} = \frac{V_1 N V_2}{2\pi f_s L_{Yd}}\left(\delta - \frac{\pi}{6}\right) \tag{2}$$

Figure 2c presents the three-phase transformer in Dy connection. This connection is simply the Yd connection operated in reversed direction. The LVS of the transformer is connected in delta (D) and the HVS of the transformer is connected in wye (y). The power transfer equation for phase-shift angles in the range of $-\frac{\pi}{3} \leq \delta \leq 0$ is,

$$P_o = P_{Dy} = \frac{V_1 N V_2}{2\pi f_s L_{Dy}}\left(\delta + \frac{\pi}{6}\right) \tag{3}$$

Since that the transformer configuration of Dy is the Yd configuration in reverse, the phase angle is also the opposite of Yd. The leakage inductance of the 3P-BIDC in Yd and Dy connection designed to operate in buck and boost mode is calculated in Equations (4) and (5) as,

$$L_{Yd} = L_Y + N^2 L_d \tag{4}$$

$$L_{Dy} = N^2 L_D + L_y \tag{5}$$

According to [15,16], the 3P-BIDC can operate under soft-switching mode when the DC-link voltage ratio x is in the range of $\frac{3}{2} \leq x \leq 2$ and $\frac{1}{2} \leq x \leq \frac{2}{3}$ for Dy and Yd configurations, respectively.

2.3. Delta-Delta (Dd) Connection

Figure 2d presents the three-phase transformer in Dd connection. Both the primary and secondary side of the transformer are connected in delta connection. The phase shift equation is shown in Equation (6) for phase-shift angles of $0 \leq \delta \leq \frac{\pi}{3}$ as,

$$P_o = P_{Dd} = \frac{V_1 N V_2}{2\pi f_s L_{Dd}} \delta\left(2 - \frac{3\delta}{2\pi}\right) \tag{6}$$

The leakage inductance, L_{Dd} of the 3P-BIDC is equal to L_{Yy}. The 3P-BIDC in Dd connection is designed to operate with a DC voltage ratio of 1:1.

3. 3P-BIDC Simulation Model and Burst-Mode Strategy

In light-load conditions, the converter is operated in such a way that the PSM is combined with burst-mode switching to generate an intermittent power transfer. In medium to high-load conditions, only the PSM is employed. The PSM strategy is a continuous switching operation of the 3P-BIDC, while the combination of PSM and burst-mode switching results in intermittent operation of the 3P-BIDC in light-load condition.

Figure 3 presents the theory of generating the burst-mode signals. Note that m is the conducting period and n is the non-conducting period of the burst-mode signal in percentage. The burst-mode strategy is generated by multiplying two signals. The continuous signal of 20 kHz is multiplied with a low frequency signal of 20 Hz. The product is an intermittent signal with a low frequency of 20 Hz. If a 50 ms low frequency signal with a duty cycle of 30% is multiplied with a train of a 50 μs signal of duty cycle 50% each, 300 gate pulses with a period of 50 μs and duty cycle 50% will last for 15 ms and there will be no gate pulses for 35 ms, the average output power will be reduced.

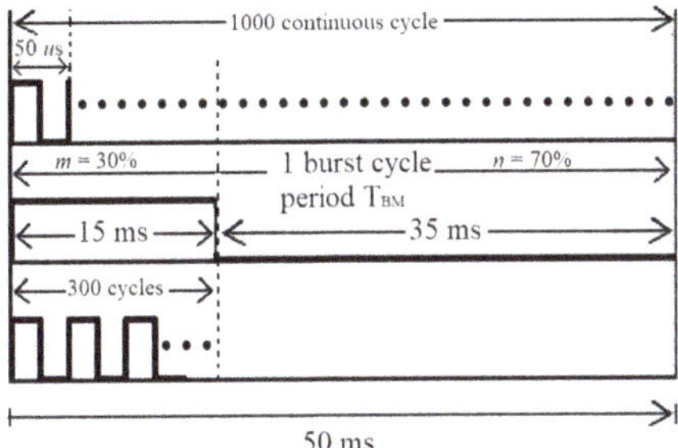

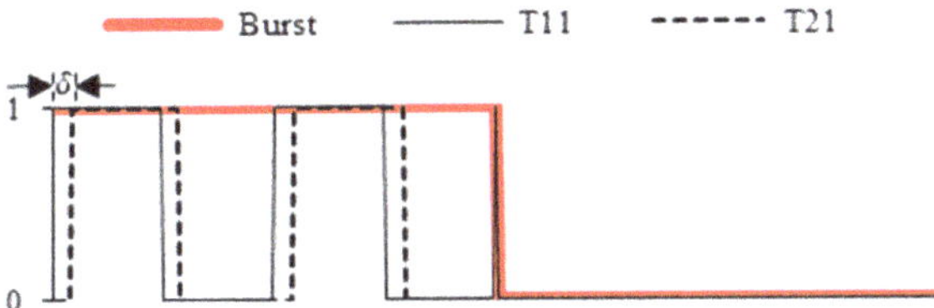

Figure 3. Generation of burst-mode signals by multiplying a continuous 20 kHz signal with one cycle of 20 Hz signal with a conduction period, *m*, of 30%.

Figure 4 presents the last two cycles of gating signals for T_{11} and T_{21} that are transitioning from the conducting period to the non-conducting period. The gating signals have a switching frequency of 20 kHz and are phase-shifted by $\pi/6$. The burst mode signal in red has a frequency of 20 Hz. Therefore, T_{11} and T_{21} will be switching at 20 kHz when the burst mode signal is in the high state. The intermittent operation is applied for power transfer less than 0.44 p.u. of the rated power.

Figure 4. Gating signals to switches T_{11} and T_{21} at $\delta = 30°$ when multiplied with the burst signal.

The phase-shift angle ranges from 0 to $\pi/6$ for power transfer from zero to the rated power using continuous operation. Moreover, with the same phase-shift angle, when intermittent operation is applied, the average DC output power is reduced.

Figure 5 shows the simulation model of the 3P-BIDC connected to a battery bank. The HF transformer is varied according to Figure 2. The simulation model considers losses such as copper and switching losses in order to represent a practical converter. The battery is modelled with an internal resistance, R_{int}, of 5 mΩ. A resistor Rs is connected in series with the transformer to represent copper loss in the transformer windings. On-state collector-emitter voltage of 1.85 V and forward voltage drop of 2.17 V are considered in the IGBTs and diodes, respectively. The IGBT model number is SKM75GB12V with maximum voltage and continuous current ratings of 1.2 kV and 114 A, respectively.

Table 1 shows the 3P-BIDC simulation model parameters. Each design differs in the value of the DC-link voltages, range of phase-shift angle, and the leakage inductances. When the converter is operated in Yd or Dy, the DC-link voltage supply that is connected to the wye side of the transformer is supplied with 520 V and the DC-link voltage that is connected to the delta side of the transformer is supplied with 300 V. This is to allow the 3P-BIDC to operate in buck and boost mode respectively. The rated power of 3 kW is designed to be achieved at $\delta = \pi/6$ for Yy and Dd connections. On the other hand, the rated power of 3 kW is achieved at $\delta = \pi/3$ for Yd and at $\delta = 0$ for Dy connections.

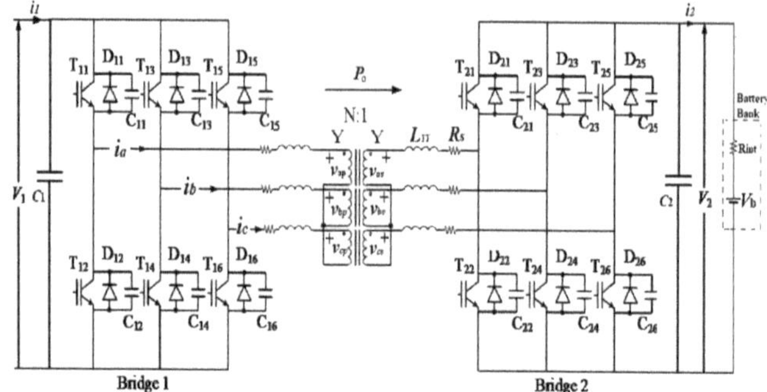

Figure 5. Simulation model of the three-phase bidirectional isolated DC-DC converter (3P-BIDC) with the Yy transformer connection.

Table 1. 3P-BIDC Circuit Parameters.

Parameters	Symbol	Values
Rated Power	P_R	3 kW
Dc-link voltage at bridge 1	V_1	300 V[u,w] 520 V[v]
Dc-link voltage at bridge 2	V_2	300 V[u,v] 520 V[w]
Range of phase-shift angle	δ	$-\frac{\pi}{3} \leq \delta \leq \frac{\pi}{3}$
Switching frequency	f_s	20 kHz
Dc-link capacitors	C_1, C_2	3 mF
Snubber capacitors	$C_{11}-C_{26}$	6 nF
Transformer turns ratio	$N{:}1$	1:1
Transformer leakage inductances/phase	L_a, L_b, L_c	36.5 µH[u] (0.15 p.u) 216 µH[v] (0.31 p.u)
	L_A, L_B, L_C	36.5 µH[u] (0.15 p.u) 216 µH[w] (0.31 p.u)
Transformer winding resistance/phase	R_s	15 mΩ (0.0005 p.u)

[u] Applies to Yy and Dd configuration, [v] Applies to Yd configuration, [w] Applies to Dy configuration. [u] Based on 300 V, 3 kW and 20 kHz. [v,w] Based on 520 V, 3 kW and 20 kHz.

Considering the power transfer from bridge 1 to 2, the input (P_i) and output (P_o) power are calculated as,

$$P_i = V_1 I_1 \tag{7}$$

and

$$P_o = V_2 I_2 \tag{8}$$

where I_1 and I_2 are the average current at bridges 1 and 2, respectively. The efficiency of the DC-DC converter is determined as the ratio of the input and output power. Note that, when the power is transferred from bridge 2 to 1, Equations (7) and (8) can be interchanged.

4. 3P-BIDC Simulation Results

4.1. Operating Waveforms

This section presents the results obtained from the simulation of the 3P-BIDC model with different transformer configurations operated in continuous and intermittent switching, in the PSCAD environment. Table 2 presents the DC-link voltage at bridges 1 and 2 with different transformer connections.

Table 2. DC-link voltages at bridge 1 and bridge 2.

Scenario	Dc-Link Voltage		Transformer Connection	
	V_1 (V)	V_2 (V)	HVS	LVS
1	300	300	Wye	Wye
2	520	300	Wye	Delta
3	300	520	Delta	Wye
4	300	300	Delta	Delta

Figure 6a–d shows the AC voltage and current waveforms of phase A in bridges 1 and 2 in Yy, Yd, Dy, and Dd transformer connections when the 3P-BIDC is operated using PSM at the rated power. In Figure 6a, the voltage levels of 200 V and 100 V at bridges 1 and 2 correspond to $v_{ap} = v_{as} = \frac{2}{3}V_1$ and $v_{ap} = v_{as} = \frac{1}{3}V_1$ in Yy transformer configuration. AC root mean square (RMS) current of 10 A flows in both the primary and secondary sides of the transformer. In Figure 6b, the voltage values of 347 V and 173 V at bridge 1 correspond to $v_{ap} = \frac{2}{3}V_1$ and $v_{ap} = \frac{1}{3}V_1$, respectively. The voltage value of 300 V at bridge 2 corresponds to $v_{AB} = V_2$. The peak AC current of 5.2 A is seen on the HVS, which is within the rated current of the converter. For the Dy connection, at the rated power of 3 kW, the voltage value of 300 V at bridge 1 corresponds to $v_{ab} = V_1$. The voltage values of 347 V and 173 V at bridge 2 correspond to $v_{as} = \frac{2}{3}V_2$ and $v_{as} = \frac{1}{3}V_2$, respectively. This results in a peak AC current value of 6.6 A at HVS, well within the rated current. The DC-link voltage ratios are 0.58 and 1.73 when the transformer is connected in Yd and Dy configurations, respectively. In Figure 6d, the voltage levels of 520 V at both bridges 1 and 2 correspond to $v_{ab} = V_1$ and $v_{AB} = V_2$.

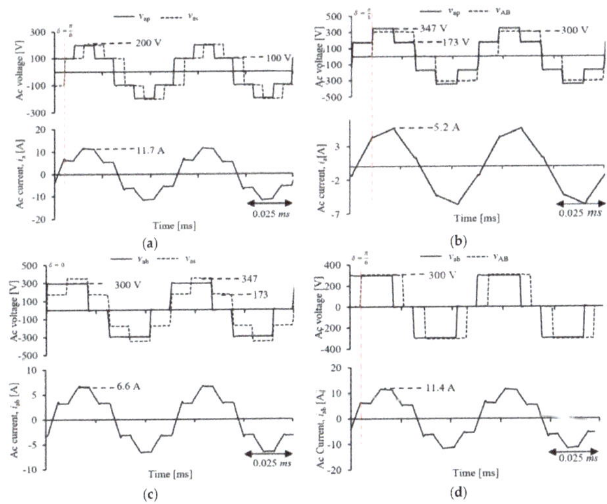

Figure 6. AC voltage and current waveforms of the 3P-BIDC in different transformer configurations operating in phase-shift modulation (PSM) at the rated power. (**a**) Yy. (**b**) Yd. (**c**) Dy. (**d**) Dd.

Figure 7a shows the AC voltage waveform of the 3P-BIDC in Yy transformer configuration for a full cycle of burst-mode with the conducting period of $m = 30\%$ and the non-conducting conducting period of $n = 70\%$ at $\delta = \frac{\pi}{6}$. Figure 7b shows the time-expanded waveform of Figure 7a from the final conducting period to the non-conducting period. During the non-conducting period, AC voltage shows a time decaying oscillation with the maximum voltage of 50 V and the rms voltage of 35.4 V.

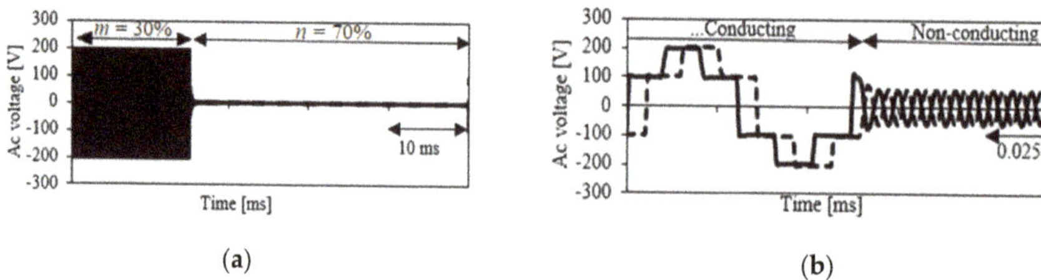

(**a**)

(**b**)

Figure 7. AC voltage waveform of the 3P-BIDC in Yy transformer connection using burst-mode. (**a**) At $\delta = \pi/6$. (**b**) Time-expanded waveform of (**a**).

4.2. Efficiency in Various Transformer Configurations

This section presents the light-load efficiency performance of the 3P-BIDC. The efficiency improvement that compares between continuous and intermittent mode is observed and evaluated at light-loads of 0.12 p.u. and 0.24 p.u. of the rated power.

Figure 8 presents the relationship between the phase-shift angles of $-\frac{\pi}{3} \leq \delta \leq \frac{\pi}{3}$ and the output power between ±1 per unit, for charging and discharging power. It can be seen that the phase-shift angle required to achieve the output power from 0 to 1 p.u. changes with different transformer connections. The power is transferred from primary side to secondary side in Yy and Dd transformer connections when $0 \leq \delta \leq \frac{\pi}{6}$. The rated power is achieved at $\delta = \frac{\pi}{6}$. In Yd and Dy configurations, the power is transferred from the primary side to the secondary side when $\frac{\pi}{6} \leq \delta \leq \frac{\pi}{3}$ and $-\frac{\pi}{6} \leq \delta \leq 0$, respectively.

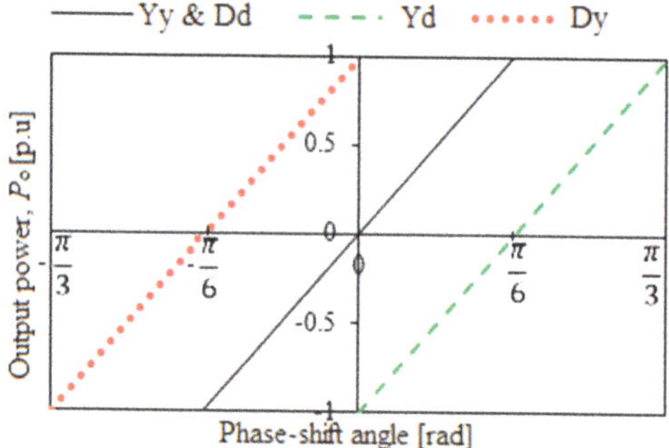

Figure 8. Phase-shift angle versus output power curve of the 3P-BIDC with different transformer configurations.

Figure 9 presents the efficiency versus output power curve of the 3P-BIDC with various transformer configurations in the continuous operation. The 3P-BIDC in Yd configuration achieved higher efficiency below 0.28 p.u. and above 0.81 p.u. of the rated power as compared to the other transformer configurations. The figure also shows that at 0.12 p.u. the efficiency of 3P-BIDC in Yy configuration is 83.6% and in Yd configuration it is 89.4%. Moreover, at 0.20 p.u., the efficiency of the 3P-BIDC in Yy configuration is 90% and in Yd configuration it is 93%. There is a significant drop in efficiency in the 3P-BIDC during light-load operation (<0.3 p.u.) for all transformer configurations. The 3P-BIDC has poor efficiency when connected to the Dy transformer.

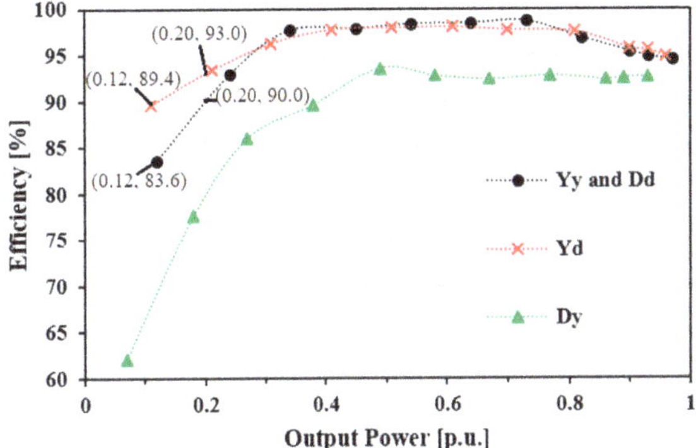

Figure 9. Efficiency versus output power curve of the 3P-BIDC with different transformer configurations in continuous operation.

Figure 10 presents the improvement in 3P-BIDC efficiency of different transformer configurations when intermittent switching is employed. Figure 10a shows that when the 3P-BIDC is connected in Yy or Dd transformer configuration, higher efficiency is achieved with $m = 30\%$ compared to $m = 10\%$ from power 0.12 p.u. to 0.3 p.u. in terms of improving the converter efficiency. An efficiency as high as 99.6% is achieved at the power transfer of 0.29 p.u when intermittent operation is employed with $m = 30\%$. Figure 10b shows that in Yd transformer configuration, the 3P-BIDC efficiency using intermittent switching with $m = 10\%$ is higher compared to $m = 30\%$ from power transfer of 0.12 p.u. to 0.21 p.u. For example, at the power transfer of 0.16 p.u., the efficiency of the 3P-BIDC with intermittent switching $m = 10\%$ is 96%, whereas the efficiency of the 3P-BIDC with $m = 30\%$ is 94.1%. At the power transfer of 0.22 p.u., it is seen that the efficiency of the 3P-BIDC with $m = 30\%$ outperforms the efficiency of the 3P-BIDC with $m = 10\%$. At the power transfer of 0.29 p.u., the efficiency of the 3P-BIDC with $m = 10\%$ is 95.7%, whereas the efficiency of the 3P-BIDC with $m = 30\%$ is 97%. Figure 10c shows that the 3P-BIDC in Dy transformer winding configuration obtained higher efficiency improvements with $m = 10\%$ at the power transfer of 0.03 p.u. to 0.25 p.u. For example, at the power transfer of 0.18 p.u., the efficiency of the 3P-BIDC with $m = 10\%$ is 90.3%. At the power transfer of 0.25 p.u., $m = 10\%$ reached the maximum range of $\delta = \frac{\pi}{3}$ and the intermittent switching is operated with $m = 30\%$. The overall efficiency of the DC-DC converter in Dy transformer configuration is low compared to the efficiency performance of the DC-DC converter in Yy, Dd, and Yd transformer configurations, which are above 95% in the light-load conditions. The authors of [6] showed that no improvement in efficiency of the 3P-BIDC is achieved when the converter is operated in intermittent switching for medium load. However, the simulation results in Figure 10 show that the efficiency of the 3P-BIDC significantly improved during light-load operation with intermittent switching. Therefore, the intermittent switching is suitable for light-load operation of the 3P-BIDC.

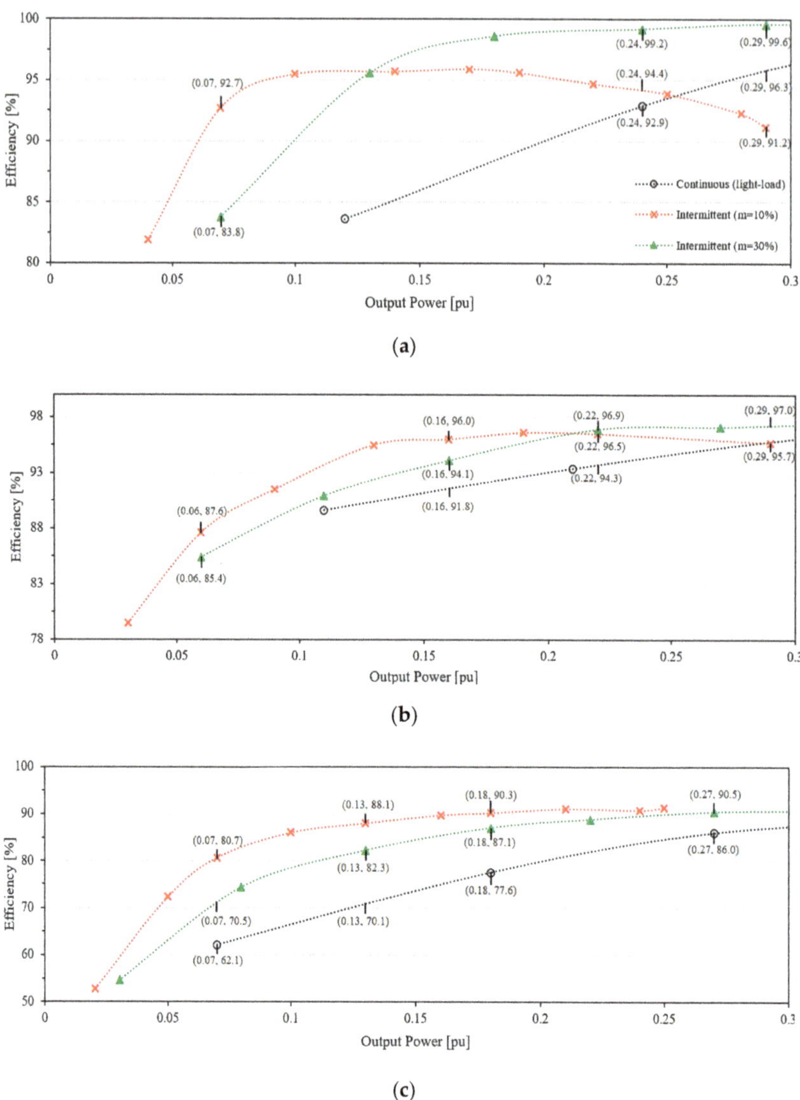

Figure 10. Efficiency versus output power curve of the 3P-BIDC with different transformer configurations in light-load conditions of less than 0.3 p.u. (**a**) Yy and Dd. (**b**) Yd. (**c**) Dy.

Tables 3 and 4 summarizes the efficiency improvement and power loss reduction of the 3P-BIDC with Yy, Yd, Dy, and Dd transformer configurations that are observed at light-load power transfers of 0.12 p.u., 0.24 p.u., and 0.29 p.u, accordingly. As shown in Table 3, the efficiency improvements of the 3P-BIDC achieved in Yy and Dd transformer configurations at light-load power transfers of 0.12 p.u, 0.24 p.u, and 0.29 p.u are higher compared to the efficiency improvement achieved in Yd transformer configuration. The DC-DC converter in Dy transformer configuration achieved high efficiency improvements. However, the overall efficiency remained below 95% in intermittent switching. Table 4 is consistent in showing that the DC-DC converter achieved the highest power loss reduction in Dy transformer configuration at light loads. However, the power loss of the DC-DC converter in

Dy transformer configuration when in intermittent operation is higher than the power losses of the DC-DC converter in Yy, Dd, and Yd transformer configurations.

Table 3. Efficiency improvement with continuous and intermittent operation in different transformer configurations.

P_o (p.u)	Transformer Configurations								
	Yy and Dd			Yd			Dy		
	Efficiency (%)		E_I (%)	Efficiency (%)		E_I (%)	Efficiency (%)		E_I (%)
	A	B		A	B		A	B	
0.12	83.6	95.6 (m = 30%)	12	89.6	93.5 (m = 10%)	3.9	67.0	87.0 (m = 10%)	20
0.24	92.6	99.2 (m = 30%)	6.6	94.9	97.0 (m = 30%)	2.1	81.8	90.9 (m = 10%)	9.1
0.29	96.0	99.6 (m = 30%)	3.6	96.0	97.3 (m = 30%)	1.3	87.8	91.2 (m = 30%)	3.4

A—Continuous mode, B—Intermittent mode, E_I (%)—Efficiency improvement as a percentage.

Table 4. Power loss reduction with continuous and intermittent operation in different transformer configurations.

P_o (p.u)	Transformer Configurations								
	Yy and Dd			Yd			Dy		
	Power Loss (p.u.)		P_{LR} (p.u.)	Power Loss (p.u.)		P_{LR} (p.u.)	Power Loss (p.u.)		P_{LR} (p.u.)
	A	B		A	B		A	B	
0.12	0.024	0.006 (m = 30%)	0.018	0.014	0.010 (m = 10%)	0.004	0.049	0.02 (m = 10%)	0.029
0.24	0.018	0.002 (m = 30%)	0.016	0.014	0.008 (m = 30%)	0.006	0.048	0.02 (m = 10%)	0.028
0.29	0.013	0.001 (m = 30%)	0.012	0.014	0.008 (m = 30%)	0.006	0.044	0.03 (m = 30%)	0.014

A—Continuous mode, B—Intermittent mode, P_{LR} (p.u)—Power loss reduction in per unit.

The 3P-BIDC has the best performance when operated with Yy and Dd transformer configuration with a DC voltage ratio that is equal to the transformer turns ratio. The Yd transformer configuration may also be suitable for applications with different DC voltage ratios such as 520 V and 300 V. On the other hand, the Dy transformer configuration is unfavorable due to the efficiency achievements of lower than 95% despite intermittent switching being applied. Note that this paper has not observed the efficiency performance of the 3P-BIDC when the DC voltage ratio of the 3P-BIDC is not the same as the transformer turns ratio.

4.3. Analysis of Current Stress in Transformer and IGBT Switches

A stress analysis was conducted on the modelled three-phase transformer and switches to compare the amount of current stresses on each of the different transformer configurations with the turns ratio of 1:1. This section theoretically analyses the current stress in the transformer and switches of 3P-BIDC in different transformer configurations, based on the method in [15]. This theoretical analysis is then compared with the simulation results.

Table 5 shows the equations used to theoretically calculate the transformer rms current and the switch rms and transient currents in the 3P-BIDC in the various transformer configurations. The variable d represents the voltage conversion ratio. In this analysis, d is always equal to 1. This shows that the voltage conversion ratio is equal to the transformer turns ratio in all transformer configurations.

Table 5. Transformers RMS current and switch RMS and transient currents.

Transformer Configurations	Range of Phase Shift Angle (rad)	$I_{\mathrm{T_{rms}}}$	$\sum I_{\mathrm{rms}}$	$\sum I_{\mathrm{sw}}$
Yy	$0 \le \delta \le \frac{\pi}{3}$	$\dfrac{V_1}{\sqrt{243}\pi\omega L_{\mathrm{Yy}}}r_1$	$\dfrac{2V_1}{\sqrt{243}\pi\omega L_{\mathrm{Yy}}}r_1$	$\left\lvert\dfrac{2V_1}{9\omega L_{\mathrm{Yy}}}\right\rvert p_1$
Yd	$0 \le \delta \le \frac{\pi}{3}$	$\dfrac{V_1}{\sqrt{243}\omega L_{\mathrm{Yd}}}m_1$	$\dfrac{\left(\sqrt{3}+3\right)V_1}{27\omega L_{\mathrm{Yd}}}m_1$	$\left\lvert\dfrac{2V_1}{3\omega L_{\mathrm{Yd}}}\right\rvert g_1$
Dy	$-\frac{\pi}{3} \le \delta \le 0$	$\dfrac{V_1}{\sqrt{243}\pi\omega L_{\mathrm{Dy}}}j_1$	$\dfrac{\left(3+\sqrt{3}\right)V_1}{27\omega L_{\mathrm{Dy}}}j_1$	$\left\lvert\dfrac{2V_1}{3\omega L_{\mathrm{Dy}}}\right\rvert q_1$
Dd	$0 \le \delta \le \frac{\pi}{3}$	$\dfrac{V_1}{9\sqrt{\pi}\omega L_{\mathrm{Dd}}}r_1$	$\dfrac{6V_1}{\sqrt{243}\pi\omega L_{\mathrm{Dd}}}r_1$	$\left\lvert\dfrac{2V_1}{3\omega L_{\mathrm{Dd}}}\right\rvert p_1$

Figure 11 presents the different current stresses in the transformers and switches versus the output average current of the 3P-BIDC in per unit terms based on the converter rated current. In Yy, Dd, and Dy, the base current is 10 A and in Yd, the base current is 6 A. The base current is multiplied by 2 for analysis that involves summation of currents in both sides of the converter.

This analysis is conducted for the power transfer range of 0 to the rated power, where the average current at the DC side ranges between 0 and 1 p.u. Figure 11a shows the rms phase current of the transformer. It is shown that the current stress increases significantly with the output average current in Dd transformer configuration. It exceeds the rated current at the output average current of 0.76 p.u. Figure 11b shows the summation of the rms current of one phase in bridges 1 and 2, which is used to determine the conduction current stress on the semiconductor switches. The rms current stress of the switches exceeds 1 p.u. from very low output current for Yd, and from output current of 0.42 p.u. for Dd transformer configurations.

The Yy transformer configuration results in lowest conduction current stress on the switches. Figure 11c shows the summation of the current of one phase in bridges 1 and 2 during a switching instant. The Yd and Dy transformer configurations are the least sensitive to the changes in output average current. However, the rms current stresses in the transient modes of the 3P-BIDC in Yd, Dy, and Dd transformer configurations exceed 1 p.u. for all output current range, which indicates high switching current stress for those transformer configurations. The switching current stress in Yy transformer configuration is the lowest of all the four types of transformer configurations and does not exceed the rated current except only after 0.82 p.u. of average output current. Therefore, it can be seen that the Yy transformer configuration is most suitable for 3P-BIDC in terms of low current stress when d is 1.

where

$$r_1 = \sqrt{\pi^3\left(5d^2 - 10d + 5\right) + d\left(-27\delta^3\right) + 54\delta^2\pi}, \quad p_1 = \left\lvert 2\pi + d(3\delta - 2\pi)\right\rvert + \left\lvert 3\delta + 2\pi(d-1)\right\rvert,$$

$$m_1 = \sqrt{\pi^2\left(15d^2 - 15d + 5\right) + d\left(81\delta^2 - 27\delta\pi\right)}, \quad g_1 = \pi\left(\left\lvert d - \frac{2}{3}\right\rvert + \left\lvert 2d - 1\right\rvert\right),$$

$$j_1 = \sqrt{\pi^2\left(5d^2 - 15d + 15\right) + d\left(81\delta^2 + 27\delta\pi\right)}, \quad q_1 = \pi\left(\left\lvert d - 2\right\rvert + \left\lvert \frac{2d}{3} - 1\right\rvert\right).$$

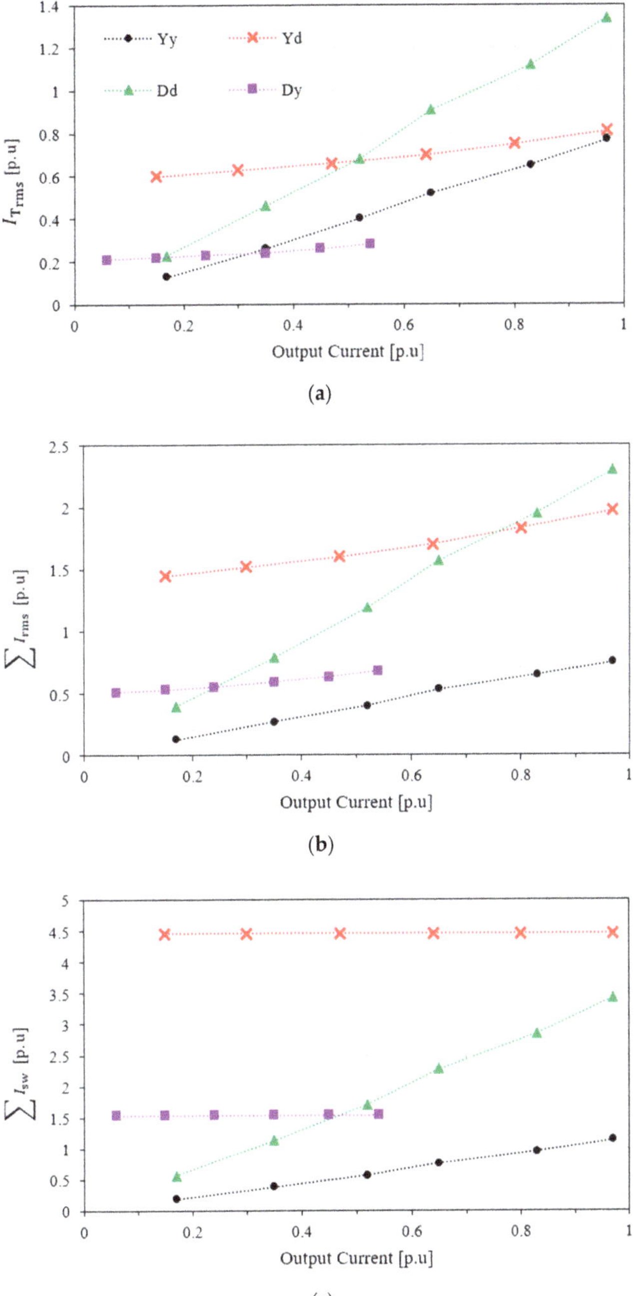

Figure 11. Current stresses in 3P-BIDC. (**a**) Transformer rms phase current. (**b**) Summation of the rms currents in one phase of bridges 1 and 2. (**c**) Summation of currents at switching instants in one phase of bridges 1 and 2.

Figure 12 presents the comparison of the theoretical analysis and the simulated analysis of the 3P-BIDC in Yy transformer configuration. In Figure 12a, the stress analysis in the rms current of the transformer in the simulation differs only by 1% to the theoretical results. Figure 12b shows that the highest error percentage of the analysis is 8% by comparing the summation of the rms currents in the switches that corresponds to one phase of bridges 1 and 2.

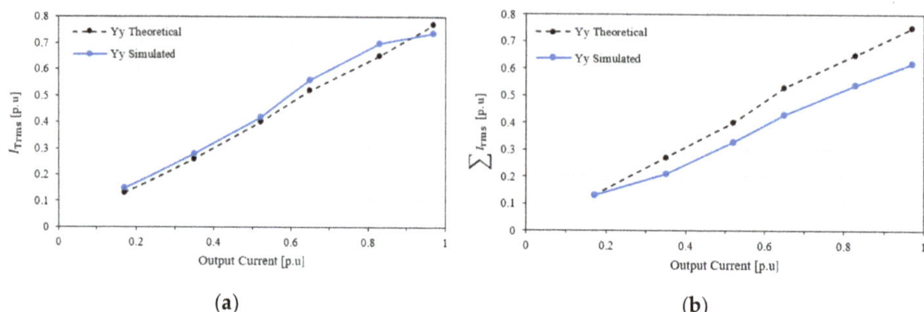

Figure 12. Comparison of theoretical and simulation results of current stress analysis. (**a**) Transformer rms current. (**b**) Summation of rms currents in one phase of bridges 1 and 2 of the 3P-BIDC.

5. Power Loss Estimation

This section presents the estimated power loss distribution and calculation details of the 3P-BIDC at light-load conditions of 0.12 p.u. and 0.34 p.u. of the rated power in the continuous mode of operation and comparing it with the light-load conditions of 0.12 p.u. and 0.34 p.u. of the rated power with a Yy transformer connection. The types of losses considered in the simulation model are copper losses, conduction losses, and snubber losses. Note that snubber losses are not considered in the power loss calculations in intermittent operation.

5.1. Distribution of Losses

5.1.1. Copper Loss

The transformer winding resistance is modelled as 15 mΩ in each phase of the transformer. The winding resistance give rise to practical copper loss in the windings of the transformer. The total copper loss is calculated as,

$$P_{cu} = 3\left[I_1{}^2 R_1 + I_2{}^2 R_2\right] \tag{9}$$

The rms current across the LVS, I_1 and the rated current across the HVS, I_2 at 0.12 p.u is 1.2 A. Therefore, the total copper loss is 0.13 W.

5.1.2. Conduction Loss

Conduction loss includes loss during the conduction of an IGBT and a diode of the model SKM75GB12V. In every switching cycle, only three switches and three diodes conduct. The conduction loss of the 3P-BIDC is calculated as,

$$P_{cond} = 3(V_{CE} + V_F)\left(I_{avg}\right) \tag{10}$$

where V_{CE} is the on-state collector-emitter voltage of the conducting IGBT, V_F is the forward voltage drop of the conducting diode, and I_{avg} is the average current.

5.1.3. Snubber Loss

The 3P-BIDC is assumed to have snubber loss if zero-voltage switching (ZVS) is not achieved. The snubber loss is calculated as,

$$P_{\text{snub}} = X C_{\text{s}} V_2{}^2 f_{\text{s}} \tag{11}$$

where X is the number of switches that is involved in snubber loss and V_2 is the DC-link voltage at the secondary side.

The converter operated at 0.12 p.u. is assumed to have hard-switching in bridge 1. Therefore, the number of switches that experience snubber loss, X, is 6. Since ZVS occurs at 0.34 p.u., the snubber loss is neglected.

5.1.4. Switching Loss

In practical conditions, the turn-off switching loss is not negligible. If ZVS turn on is achieved, turn-on switching loss is negligible. The average turn-off switching loss is directly proportional to the square of the switching current and the switch current fall time [29],

$$P_{\text{SW}} = \frac{T_f}{48 C_{\text{s}}} I_{\text{SW}}{}^2 \tag{12}$$

where I_{SW} is the turn-off switching current and T_f is the switching current fall time. However, in the simulation, only on-state voltage drop and diode forward voltage drop are considered. Therefore, T_f is very short, rendering the switching loss negligible.

5.1.5. Total Power Loss

The total power loss P_{Loss} in the 3P-BIDC is calculated as the summation of copper P_{cu}, conduction P_{cond}, and snubber P_{snub} losses,

$$P_{\text{Loss}} = P_{\text{cu}} + P_{\text{cond}} + P_{\text{snub}} \tag{13}$$

The total estimated power loss at 0.12 p.u. and 0.34 p.u. at the rated power is 74.7 W (0.025 p.u.) and 25.3 W (0.008 p.u.), respectively.

5.2. Numerical Calculation of Losses

This sub-section provides the numerical calculation of losses in the 3P-BIDC at the power transfer of 0.12 p.u. and 0.34 p.u. at the rated voltage. The power loss is calculated when only PSM and intermittent operation are employed. From Figure 10, it can be seen that the onset of ZVS is around 0.34 p.u. of power transfer. Hence, no snubber loss is accounted for at the power transfer of 0.34 p.u.

5.2.1. Power Loss at 0.12 p.u ($\delta = 4°$) in Continuous Mode

- Copper loss

$$\begin{aligned}
P_{\text{cu}} &= 3\left[I_1{}^2 R_1 + I_2{}^2 R_2\right] \\
P_{\text{cu}} &= 3\left[1.2^2(0.015) + 1.2^2(0.015)\right] \\
&= 0.13 \text{ W } (0.000043 \text{ p.u.})
\end{aligned}$$

- Conduction loss

$$\begin{aligned}
P_{\text{cond}} &= 3(V_{\text{CE}} + V_{\text{F}})\left(I_{\text{avg}}\right) \\
P_{\text{cond}} &= 3(1.85 + 2.17)(0.816) \\
&= 9.8 \text{ W } (0.0033 \text{ p.u.})
\end{aligned}$$

- Snubber loss

$$P_{snub} = XC_s V_b^2 f_s$$
$$P_{snub} = (6)\left(6 \times 10^{-9}\right)(300)^2\left(20 \times 10^3\right)$$
$$= 64.8 \text{ W } (0.0216 \text{ p.u.})$$
$$P_{Loss} = P_{cu} + P_{cond} + P_{snub}$$
$$= 0.13 \text{ W} + 9.8 \text{ W} + 64.8 \text{ W}$$
$$= 74.73 \text{ W } (0.025 \text{ p.u.})$$

5.2.2. Power Loss at 0.34 p.u. ($\delta = 10°$) in Continuous Mode

- Copper loss

$$P_{cu} = 3\left[I_1^2 R_1 + I_2^2 R_2\right]$$
$$P_{cu} = 3\left[3.63^2(0.015) + 3.63^2(0.015)\right]$$
$$= 1.19 \text{ W } (0.0004 \text{ p.u.})$$

- Conduction loss

$$P_{cond} = 3(V_{CE} + V_F)\left(I_{avg}\right)$$
$$P_{cond} = 3(1.85 + 2.17)(2)$$
$$= 24.12 \text{ W } (0.008 \text{ p.u.})$$
$$P_{Loss} = P_{cu} + P_{cond} + P_{snub}$$
$$= 1.19 \text{ W } + 24.12 \text{ W } + 0 \text{ W}$$
$$= 25.31 \text{ W } (0.008 \text{ p.u.})$$

The losses in intermittent operation are calculated with the same equation as in continuous operation. However, since that this intermittent operation is carried out with $m = 30\%$, the calculated total power loss is multiplied by a factor of 0.3. The phase-shift angles to achieve the power transfer 0.12 p.u. and 0.34 p.u. in the intermittent operation are higher than when only continuous operation is employed.

5.2.3. Power Loss at 0.12 p.u. ($\delta = 7°$) in Intermittent Mode

- Copper loss

$$P_{cu} = 3\left[I_1^2 R_1 + I_2^2 R_2\right]$$
$$P_{cu} = 3\left[1.18^2(0.015) + 1.18^2(0.015)\right]$$
$$= 0.125 \text{ W } (0.000042 \text{ p.u.})$$

- Conduction loss

$$P_{cond} = 3(V_{CE} + V_F)\left(I_{avg}\right)$$
$$P_{cond} = 3(1.85 + 2.17)(1.4)$$
$$= 16.9 \text{ W } (0.006 \text{ p.u.})$$

- Snubber loss

$$P_{\text{snub}} = XC_s V_b{}^2 f_s$$
$$P_{\text{snub}} = (6)\left(6 \times 10^{-9}\right)(300)^2\left(20 \times 10^3\right)$$
$$= 64.8 \text{ W } (0.0216 \text{ p.u.})$$
$$P_{\text{Loss}} = P_{\text{cu}} + P_{\text{cond}} + P_{\text{snub}}$$
$$= 0.13 \text{ W } + 9.8 \text{ W } + 64.8 \text{ W}$$
$$= 74.73 \text{ W } (0.025 \text{ p.u})$$
$$P_{\text{Loss}} = 0.3\left[P_{\text{cu}} + P_{\text{cond}} + P_{\text{snub}}\right]$$
$$= 0.3\left[1.125 \text{ W } + 16.9 \text{ W } + 64.8 \text{ W}\right]$$
$$= 24.5 \text{ W } (0.008 \text{ p.u.})$$

5.2.4. Power Loss at 0.34 p.u. ($\delta = 19°$) in Intermittent Mode

- Copper loss

$$P_{\text{cu}} = 3\left[I_1{}^2 R_1 + I_2{}^2 R_2\right]$$
$$P_{\text{cu}} = 3\left[4.9^2(0.015) + 4.9^2(0.015)\right]$$
$$= 2.16 \text{ W } (0.00072 \text{ p.u.})$$

- Conduction loss

$$P_{\text{cond}} = 3(V_{\text{CE}} + V_{\text{F}})\left(I_{\text{avg}}\right)$$
$$P_{\text{cond}} = 3(1.85 + 2.17)(2.8)$$
$$= 33.8 \text{ W } (0.0071 \text{ p.u.})$$
$$P_{\text{Loss}} = 0.3\left[P_{\text{cu}} + P_{\text{cond}} + P_{\text{snub}}\right]$$
$$= 0.3\left[2.16 \text{ W } + 33.8 \text{ W } + 0 \text{ W}\right]$$
$$= 10.77 \text{ W } (0.004 \text{ p.u.})$$

Considering the Yy transformer winding configuration, Figure 13 presents the calculated total loss, compared with the total loss measured from the simulation model, in continuous and intermittent mode of operations focusing on the power levels 0.12 p.u. to 0.34 p.u. The figure shows that the minimum power loss is obtained at the power transfer of 1.02 kW, which also indicates the onset of ZVS [19], hence snubber loss can be negligible.

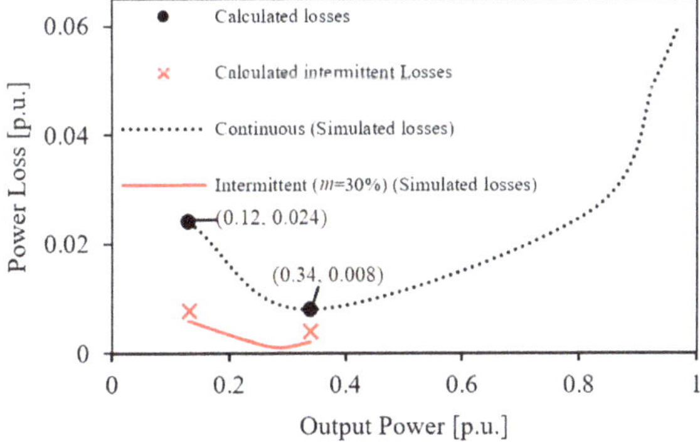

Figure 13. Output power versus power loss of the 3P-BIDC in continuous mode and intermittent mode during light-load condition with Yy transformer configuration.

Figure 14a,b presents the copper, conduction, and snubber losses in per unit terms in the continuous mode of operation. The copper loss increases as the output power increases. This is because there is a higher rated current flow through the transformer, resulting in higher I^2R losses.

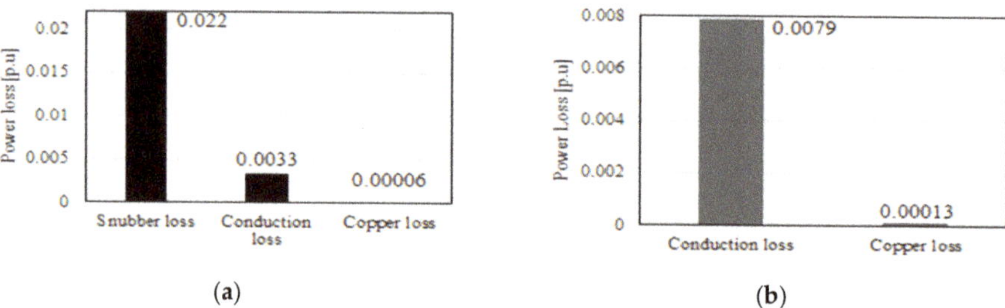

(a)   (b)

Figure 14. Power loss distribution of the 3P-BIDC in continuous operation; snubber, copper, and conduction losses at (**a**) 0.12 p.u. (**b**) 0.34 p.u.

Figure 15a,b presents the copper and conduction per unit losses in intermittent operation. It is shown that the intermittent operation applied at 0.12 p.u. and 0.34 p.u. reduced the amount of losses by 3–4%. This is because the number of switching cycles are reduced, therefore reducing the total losses in the output.

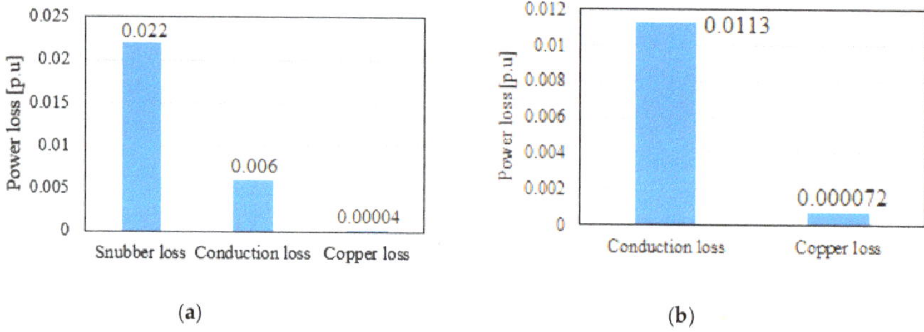

(a)   (b)

Figure 15. Power loss distribution of the 3P-BIDC in intermittent operation; copper loss and conduction loss at (**a**) 0.12 p.u. (**b**) 0.34 p.u.

6. Conclusions

This paper presented the operational feasibility and efficiency performance evaluation of the 3 kW 3P-BIDC in Yy, Yd, Dy, or Dd transformer winding configuration using the combination of PSM and burst-mode switching to improve the low power transfer efficiency of the converter. The 3P-BIDC achieved the highest efficiency performance in Yy and Dd transformer configurations in light-load power transfers in intermittent operation. However, the Yd transformer configuration is suitable and can result in high efficiency when the DC voltage is not equal on either side of the 3P-BIDC. It is not preferred to operate the 3P-BIDC with a Dy transformer connection as it results in the overall poorest 3P-BIDC efficiency performance amongst the other transformer configurations. The theoretical current stress analysis shows that the 3P-BIDC operated with Yy transformer configuration results in the lowest current stress in the transformer and switches. Moreover, the loss analysis is comparable to the stress and loss measured in the simulation model for the 3P-BIDC in Yy transformer configuration.

Author Contributions: Conceptualization, N.M.L.T.; methodology, N.S.M.S. and N.M.L.T.; simulation, N.S.M.S.; validation, N.S.M.S., N.M.L.T. and H.A.; formal analysis, N.S.M.S. and N.M.L.T.; investigation, N.S.M.S.; resources, N.S.M.S. and N.M.L.T.; data curation, N.S.M.S.; writing—original draft preparation, N.S.M.S. and N.M.L.T.; writing—review and editing, N.S.M.S., N.T. and H.A.; visualization, N.M.L.T.; supervision, N.M.L.T.; project administration, N.M.L.T.; funding acquisition, N.M.L.T. All authors have read and agreed to the published version of the manuscript.

Funding: This research was financially supported by the Ministry of Higher Education through FRGS Project No. FRGS/1/2017/TK04/UNITEN/02/4. The APC was supported by UNITEN BOLD Publication Fund.

Conflicts of Interest: The authors declare no conflict of interest.

References

1. De Doncker, R.W.A.A.; Divan, D.M.; Kheraluwala, M.H. A Three-Phase Soft-Switched High-Power-Density DC/DC Converter for High-Power Applications. *IEEE Trans. Ind. Appl.* **1991**, *27*, 63–73. [CrossRef]
2. Tan, N.M.L.; Inoue, S.; Kobayashi, A.; Akagi, H. An energy storage system combining a 320-V, 12-F electric double layer capacitor bank with a bidirectional isolated DC-DC converter. *IEEE Trans. Power Electron.* **2008**, *23*, 2755–2765. [CrossRef]
3. Nguyen, D.D.; Bui, N.T.; Yukita, K. Design and optimization of three-phase dual-active-bridge converters for electric vehicle charging stations. *Energies* **2019**, *13*, 150. [CrossRef]
4. Lei, T.; Wu, C.; Liu, X. Multi-objective optimization control for the aerospace dual-active bridge power converter. *Energies* **2018**, *11*, 1168. [CrossRef]
5. Tan, N.M.L.; Abe, T.; Akagi, H. Design and performance of a bidirectional isolated DC-DC converter for a battery energy storage system. *IEEE Trans. Power Electron.* **2012**, *27*, 1237–1248. [CrossRef]
6. Baars, N.H.; Everts, J.; Huisman, H.; Duarte, J.L.; Lomonova, E.A. A 80-kW Isolated DC–DC Converter for Railway Applications. *IEEE Trans. Power Electron.* **2015**, *30*, 6639–6647. [CrossRef]
7. De Din, E.; Siddique, H.A.B.; Cupelli, M.; Monti, A.; De Doncker, R.W. Voltage Control of Parallel-Connected Dual-Active Bridge Converters for Shipboard Applications. *IEEE J. Emerg. Sel. Top. Power Electron.* **2018**, *6*, 664–673. [CrossRef]
8. Saeed, M.; Rogina, M.R.; Rodríguez, A.; Arias, M.; Briz, F. SiC-Based High Efficiency High Isolation Dual Active Bridge Converter for a Power Electronic Transformer. *Energies* **2020**, *13*, 1198. [CrossRef]
9. Sathishkumar, P.; Krishna, T.N.; Khan, M.A.; Zeb, K.; Kim, H.-J. Digital Soft Start Implementation for Minimizing Start-Up Transients in High Power DAB-IBDC Converter. *Energies* **2018**, *11*, 956. [CrossRef]
10. Calderon, C.; Barrado, A.; Rodriguez, A.; Alou, P.; Lazaro, A.; Fernandez, C.; Zumel, P. General Analysis of Switching Modes in a Dual Active Bridge with Triple Phase Shift Modulation. *Energies* **2018**, *11*, 2419. [CrossRef]
11. Krismer, F.; Kolar, J.W. Closed form solution for minimum conduction loss modulation of DAB converters. *IEEE Trans. Power Electron.* **2012**, *27*, 174–188. [CrossRef]
12. Everts, J. Closed-Form Solution for Efficient ZVS Modulation of DAB Converters. *IEEE Trans. Power Electron.* **2017**, *32*, 7561–7576. [CrossRef]
13. Oggier, G.G.; Ordonez, M. High-efficiency DAB converter using switching sequences and burst mode. *IEEE Trans. Power Electron.* **2016**, *31*, 2069–2082. [CrossRef]
14. Yan, G.; Li, Y.; Jia, Q. Comparative Analysis of Single and Three-phase Dual Active Bridge Bidirectional DC-DC Converter Based on the Phase-Shifting Control. In Proceedings of the 2016 International Conference on Energy, Power and Electrical Engineering (EPEE 2016), Shenzhen, China, 30–31 October 2016; pp. 326–333.
15. Núñez, R.O.; Oggier, G.G.; Botterón, F.; García, G.O. A comparative study of Three–Phase Dual Active Bridge Converters for renewable energy applications. *Sustain. Energy Technol. Assess.* **2017**, *23*, 1–10. [CrossRef]
16. Baars, N.H.; Everts, J.; Wijnands, C.G.E.; Lomonova, E.A. Performance Evaluation of a Three-Phase Dual Active Bridge DC-DC Converter with Different Transformer Winding Configurations. *IEEE Trans. Power Electron.* **2016**, *31*, 6814–6823. [CrossRef]
17. Sharifuddin, N.S.M.; Tan, N.M.L.; Akagi, H. Low-load Efficiency Improvement of a Three-Phase Bidirectional Isolated DC-DC Converter (3P-BIDC) Via Enhanced Switching Strategy. *Int. J. Eng. Technol.* **2018**, *7*, 932–938. [CrossRef]

18. Van Hoek, H.; Jacobs, K.; Neubert, M.; De Doncker, R.W. Enhanced operating strategy for a three-phase dual-active-bridge converter including frequency variation. In Proceedings of the 2015 IEEE 11th International Conference on Power Electronics and Drive Systems, Sydney, Australia, 9–12 June 2015; pp. 492–498.
19. Choi, H.J.; Jung, S.H.; Jung, J.H. A Novel Switching Algorithm to improve Efficiency at light load conditions for Three-Phase DAB Converter in LVDC Application. In Proceedings of the 2018 International Power Electronics Conference (IPEC-Niigata 2018-ECCE Asia), Niigata, Japan, 20–24 May 2018; pp. 383–387.
20. Haneda, R.; Akagi, H. Design and Performance of the 850-V 100-kW 16-kHz Bidirectional Isolated DC-DC Converter Using SiC-MOSFET/SBD H-Bridge Modules. *IEEE Trans. Power Electron.* **2020**. Early Access. [CrossRef]
21. Tan, Z.Y.; Tan, N.M.L.; Hussain, I.S. Theoretical Analysis of a Three-Phase Bidirectional Isolated DC-DC Converter Using Phase-Shifted Modulation. *Int. J. Power Electron. Drive Syst.* **2018**, *9*, 495–503. [CrossRef]
22. Engel, S.P.; Soltau, N.; Stagge, H.; De Doncker, H. Dynamic and balanced control of three-phase high-power dual-active bridge DC-DC converters in DC-grid applications. *IEEE Trans. Power Electron.* **2013**, *28*, 1880–1889. [CrossRef]
23. Wang, B.; Xiaoni, X.; Wu, S.; Wu, H.; and Ying, J. Analysis and implementation of LLC burst mode for light load efficiency improvement. In Proceedings of the 2009 Twenty-Fourth Annual IEEE Applied Power Electronics Conference and Exposition, Washington, DC, USA, 15–19 February 2009; pp. 58–64.
24. Reverter, F.; Gasulla, M. Optimal Inductor Current in Boost DC/DC Converters Operating in Burst Mode Under Light-Load Conditions. *IEEE Trans. Power Electron.* **2016**, *31*, 15–20. [CrossRef]
25. Vasic, D.; Liu, Y.P.; Costa, F.; Schwander, D. Piezoelectric transformer-based DC/DC converter with improved burst-mode control. In Proceedings of the 2013 IEEE Energy Conversion Congress and Exposition, Denver, CO, USA, 15–19 September 2013; pp. 147–153.
26. Abramson, R.A.; Gunter, S.J.; Otten, D.M.; Afridi, K.K.; Perreault, D.J. Design and evaluation of a reconfigurable stacked active bridge dc/dc converter for efficient wide load-range operation. In Proceedings of the IEEE Applied Power Electronics Conference and Exposition (APEC), Tampa, FL, USA, 26–30 March 2017; pp. 3391–3401.
27. Trescases, O.; Wen, Y. A survey of light-load efficiency improvement techniques for low-power dc-dc converters. In Proceedings of the 8th International Conference on Power Electronics-ECCE Asia, Jeju, Korea, 30 May–3 June 2011; pp. 326–333.
28. Jang, Y.; Jovanovic, M.M.; Dillman, D.L. Light-Load Efficiency Optimization Method. In Proceedings of the Twenty-Fourth Annual IEEE Applied Power Electronics Conference and Exposition (APEC 2009), Washington, DC, USA, 15–19 February 2009; Volume 25, pp. 1138–1144.
29. Akagi, H.; Yamagishi, T.; Tan, N.M.L.; Miyazaki, Y.; Kinouchi, S.I.; Koyama, M. Power-loss breakdown of a 750-V 100-kW 20-kHz bidirectional isolated DC-DC converter using SiC-MOSFET/SBD dual modules. *IEEE Trans. Ind. Appl.* **2015**, *51*, 420–428. [CrossRef]

Article

From Non-Modular to Modular Concept of Bidirectional Buck/Boost Converter for Microgrid Applications

Michal Frivaldsky *, Slavomir Kascak, Jan Morgos and Michal Prazenica

Department of Mechatronics and Electronics, Faculty of Electrical Engineering and Information Technologies, University of Zilina, 01026 Zilina, Slovakia; slavomir.kascak@feit.uniza.sk (S.K.); jan.morgos@feit.uniza.sk (J.M.); michal.prazenica@feit.uniza.sk (M.P.)
* Correspondence: Michal.frivaldsky@fel.uniza.sk

Received: 30 May 2020; Accepted: 24 June 2020; Published: 26 June 2020

Abstract: In this article, the practical comparison of the operational performance of the modular (or multiport) and non-modular bidirectional buck/boost (bi-BB) DC/DC converter is realized. The main contribution of the work is the evaluation of both concepts based on various aspects, considering the qualitative indicators of the systems relevant for microgrids. Here, we discuss efficiency, electrical properties, costs, and component values. At the same time, critical comparisons are provided for converters based on SiC and GaN technology (non-modular high-voltage SiC-based dual-interleaved converter and modular low-voltage GaN-based). The concepts are specific with their operating frequency, whereby for each solution, the switching frequency is different and directly influences relevant components. The efficiency, overall system volume, output voltage ripple, and input current ripple are compared mutually between both concepts with a dependency on power delivery. These factors, together with overall volume and costs, are very important considering modern converters for microgrid systems. The summary of pros and cons is realized for each of the proposed converters, whereby the evaluation criterion is reflected within the electrical properties targeting microgrid application.

Keywords: bidirectional converter; high efficiency; GaN; SiC; buck-boost converter; high switching frequency

---

1. Introduction

Electricity generation, transmission, and distribution are being revolutionized due to various economic, technical, and environmental reasons. A microgrid (MG) is among the new technologies that have attracted great attention recently. The existing centralized grid system is actively being replaced by distributed energy resources located closer to consumers to meet their requirements effectively and reliably. A microgrid is a modern distributed power system using local, sustainable power resources designed through the various smart grid in initiatives. Energy resources such as small capacity hydro units, wind turbines, and photovoltaic systems, in cooperation with energy storage systems, are within MG for electrification. Here, we discuss mainly households where grid electricity access is not simple due to poor access to remote areas or technical skills [1–4].

A DC-based microgrid is one of the proposed architectures for geographically remote users. The considered architecture can investigate the performance and feasibility of a DC-based microgrid for the small domestic area, as illustrated in Figure 1. In this model, several types of sources, such as solar energy, a wind power generator, or an energy storage system (ESS), are connected to the DC distribution node. Each energy source is connected to a common DC node through a relevant power converter. For example, a wind generator uses an AC/DC isolated PFC converter; solar panels use MPPT boost converters; and ESS is mostly equipped by bidirectional buck/boost topology to deliver energy into DC distribution bus. The DC bus is connected to different types of load, which may require power in the form of DC or AC and can be achieved by using DC/DC buck or DC/AC converters [5,6]. Decentralized energy sources can also be considered in the context of electric vehicle batteries, which can also serve as an energy storage system if needed. For such distribution and proper cooperation of individual ESSs, the network must allow bi-directional power flow between them and even between them and DC bus [7,8].

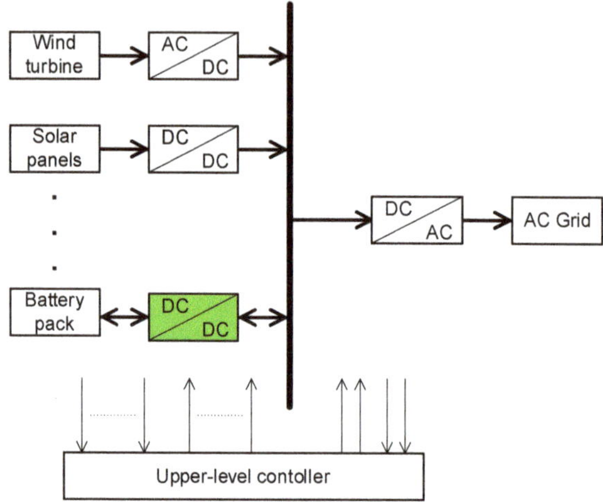

Figure 1. Typical structure of power electronics-based DC microgrid.

Bidirectional energy flow between the DC bus and ESS can be secured by a wide spectrum of power converter topologies, among which the buck/boost converter is mostly utilized due to definitions on input/output operational parameters [8–11]. Even bidirectional buck/boost (bi-BB) topology exhibits many variations (considering isolation, soft-switching, etc.); it is recommended to utilize the robust, reliable, redundant, and efficient solution. Simultaneously, such a solution shall not require high investments and cost for design and development [12–16].

Due to the mentioned fact, this paper focuses on the more detailed investigation and analyses of standard bidirectional buck/boost converter (two alternatives). At the same time, the main criterion of the evaluation is reflected within efficiency performance, costs, and input/output ripple of electrical variables. Here, an interleaved non-modular solution of a bi-BB converter equipped with SiC technology is compared to modular solution equipped by GaN transistors. Concepts differ in power semiconductor technology; thus, operating frequency and power level of individual modules comprising the whole converter system are specific for both types. For the evaluation of pros and cons, specifications on operational parameters have been defined considering DC microgrid subpart ESS–BiBB converter–DC bus. Due to purposes of laboratory testing, the converters are prototyped in a reduced power ratio, i.e., 1:10 related to power delivery and electric stress (reflected in power losses). A detailed evaluation of both technological concepts is provided, while key evaluation criteria are subjected to the main specifics of the microgrid systems (concept flexibility, complexity). The major contribution is focused on the mutual evaluation of SiC technology and GaN technology from the perspective of the application scope. In contrast, such evaluation concerning the functionality of the target application was not carried out in detail in already published studies [17–21].

2. Bi-BB Converter from Non-Modular to Modular Topology—Properties Analysis

For the purposes of the analyses related to the design of a microgrid's ESS power converter system, the focus is given on the determination of electrical properties of non-isolated bidirectional buck/boost alternative, whose principal diagram is shown on Figure 2 [4]. Since interleaved topologies are becoming increasingly utilized due to the number of positives, here we consider dual interleaved bi-BB converter as non-modular topology (Figure 3). The modular concept of n-modules will be composed of standard bi-BB converter (Figure 3), while the means of the connection of input/output terminals is described later within text.

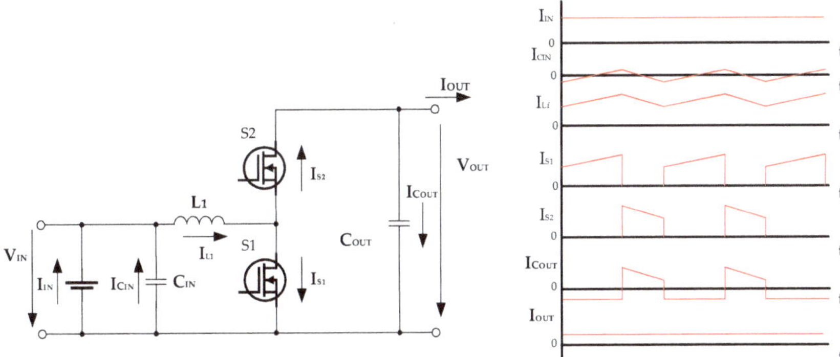

Figure 2. Operational properties of bidirectional buck/boost converter (**left**) and its circuit diagram (**right**).

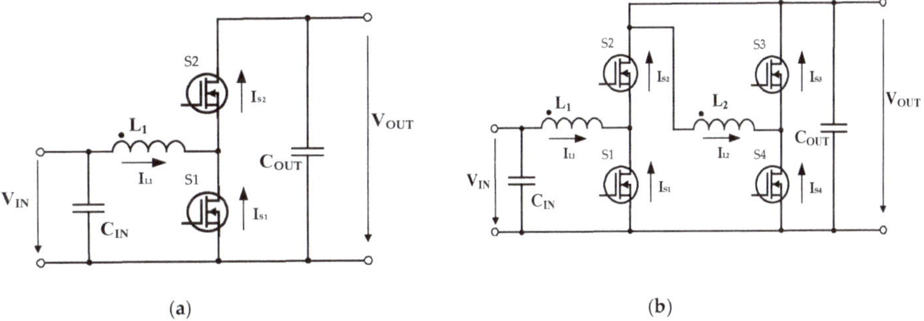

(a)   (b)

Figure 3. Circuit schematic of non-modular interleaved bi-BB converter (**a**) and circuit schematic of one module for the modular solution of bi-BB converter (**b**).

Investigation of the waveforms of voltages and currents might be considered for the output and input parts of the converter as well. Regarding current ripples, they influence the effective value of current of the output capacitor, what affects its lifetime. Therefore, analyses according to current ripples must be provided if the optimized operation of the converter is the target [22–24].

Current/Voltage Ripple Dependencies

One module of the modular converter and one two-phase non-modular converter is depicted in Figure 3. The difference in the modular and non-modular converter is that the modular converter has modules connected in series, i.e., outputs are connected in series, and input sources are independent of each other. On the other side, the connection of modules in a non-modular converter is in parallel.

The assumption is that the inductor current is continuous in cases where the converter works in a boost or a buck mode, as shown in Figure 2. In a steady state, the inductor current is the sum of the DC and AC parts. If the condition of minimal value of input capacitor Equation (1) is satisfied [25], the input current drawn from the battery is constant. In this case, the AC component is provided by the capacitor current. Then, the AC component is equal to the ripple of the inductor current, Equations (2) and (3) [17], which are valid for the boost and a buck converter, respectively.

$$C_{min} = \frac{\Delta I_L}{8 f_{sw} \Delta V_{CIN}}$$

(1)

$$\Delta I_{IN} = \Delta I_L = \Delta I_{C_{IN}} = \frac{V_{IN}}{L f_{sw}} D$$

(2)

$$\Delta I_{IN} = \Delta I_L = \Delta I_{C_{IN}} = \frac{V_{OUT} - V_{IN}}{L f_{sw}} D$$

(3)

where C_{min} is the minimal value of input capacitance, f_{sw} is the switching frequency, ΔV_{CIN} is the voltage ripple on the input capacitor, ΔI_{IN}, ΔI_L, and ΔI_{CIN} are the values of the ripple of the input current, the inductor and the input capacitor, respectively, V_{IN} is the input supply voltage, D is the duty cycle, and V_{OUT} is the value of the converter output voltage.

The inductor current ripple, Equation (2), is dependent on the input voltage (battery voltage), duty cycle, inductor value, and the switching frequency. A modular converter, unlike a non-modular one, has separate input sources. A non-modular converter has one input source or input source connected in series (e.g., serially connected battery packs). It means that the input voltage is much lower in the case of a modular converter, and therefore the current ripple is much smaller. This fact significantly reduces the ripple of the input current. Therefore, an inductance with a much smaller value is enough to maintain the same input ripple in comparison to the non-modular solution. For example, the n-module modular converter is used, and n-series connected battery packs are used for the non-modular case, the inductor value should be n times lower for the maintenance of the same current ripple.

The character of the input capacitor current for both converters is triangular, not impulse. Therefore, the impact on a voltage ripple is smaller than in the case of the impulse current. The input voltage ripple is dependent on an AC component of the inductor current and ESR of the battery pack and input capacitor. However, due to DC current drawn from the battery, as was mentioned earlier, and the parallel-connected battery pack with high capacity to the input capacitor, the input voltage ripple is negligible.

The topology of the non-modular converter is classical interleaved bidirectional buck/boost converter. In the case of interleaved topology, it is possible to achieve a state when the input current ripple is zero due to current ripple cancelation between parallel-connected phases [26–28]. This situation is depicted in Figure 4, [29]. The ratio between the input current and the inductor current is shown. It is an advantageous property in cases where the operation of the converter is at or around the desired duty cycle. In the case of a four-phase converter, the desired duty cycle is 0.25, 0.5, and 0.75. Ideally, the input voltage ripple is also zero or almost zero. It can be seen from Figure 4 that the ripple of the input current is smaller over the entire range of the duty cycle (ΔI_L is also inductor current of the modular converter). This statement applies under the condition mentioned above, and the input voltage is the same for the modular and non-modular converter. A more detailed explanation of a current ripple for the interleaved topologies is given in Appendix A.

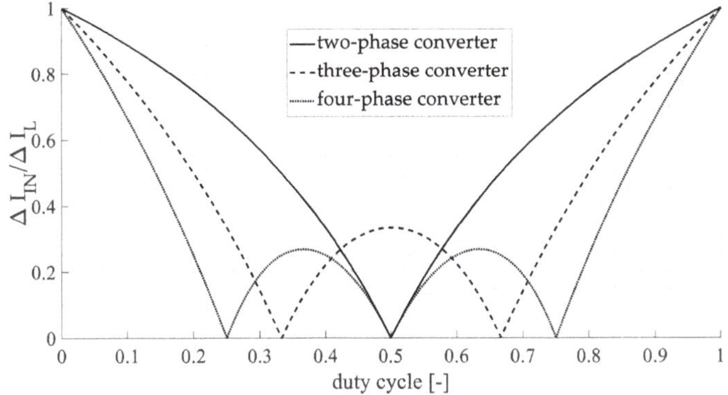

Figure 4. The input current ripple cancellation for 2-,3- and 4-phase interleaved converter, [30].

For a better image of the current and voltage ripple cancellation at the output of the modular converter, the simplification of the converter must be performed. If we simplify the given converter by the replacement of the output series-connected capacitor with one capacitor, then the current to this capacitor is continuous and does not have a pulse character (only triangular) during operation for up to 87.5% of the duty cycle. This operational mode is valid for modular concepts; thus, it meets this criterion. The value of the capacitor is then n times lower, and ESR is n times higher. The load current of the simplified converter (conventional bi-BB converter) is equal to the effective value of the capacitor current of the one module in the modular converter. The ripple of the output capacitor current is as follows:

$$\Delta I_C = \frac{V_{OUT} - V_{IN}}{8Lf_{sw}} \tag{4}$$

An explanation of the simplification and calculation of the output current ripple is given in Appendix B.

In the case of the non-modular converter, the output capacitor current is continuous when the value of the duty cycle is above 50%. Otherwise, the current has a pulse character, and the ripple is much higher. Therefore, the utilization of a modular converter is a better solution because the current ripple cancelation is within the wide operational range of the converter. The voltage ripple calculation is performed according to Equations (4) and (5), respectively [17].

$$\Delta V_C = \frac{\Delta I_C}{Cf_{sw}} \tag{5}$$

3. Bi-BB Converter Design Guideline Considering Components Selection and Costs

Since the design of the bi-BB converter must be adjusted to the nominal parameters of the target application, the input specifications are exactly defined (Table 1). The target application is considered as a low-power installation of a smart-grid node within the family house. The primary source of energy from renewable energy types is the photovoltaic block (Figure 1), which supplies the MPPT converter, whose output supplies the DC bus with 600 V of nominal voltage, which represents the input side of the bi-BB converter. For the modular system, the input voltage is divided between the serial connection of the modular converter blocks (Figure 5).

Table 1. Operational parameters of target application reflecting the situation from Figure 1.

Parameter	Value
Output voltage range from PV panels	500–560 V DC
Output power from PV panels	10 kW peak
Output voltage (DC bus voltage)	600 V DC
Output MPPT converter power	10 kW peak

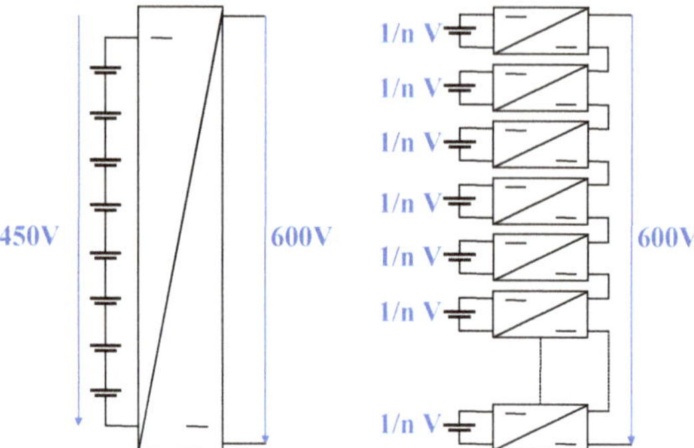

Figure 5. Block diagram of non-modular converter concept (**left**) and modular converter concept (**right**).

The input voltage for both concepts shall be 600 V, thus for the non-modular solution, the output is single. In contrast, the modular solution is characterized by the serial connection of the input terminals of the individual converters. The output of bi-BB supplies energy storage components (battery pack). At the same time, the non-modular concept is defined by 520 V of single output voltage, whereby the modular concept has an n-independent low-voltage output connected to ESS. The advantage of a modular concept is the possibility of active battery management provided by individual modules of the concept, as it has an independent output connected to batteries. It improves power management and prolongs life expectations, as discussed in [30,31]. The non-modular solution shall be equipped by additional active/passive balancing units if required.

Focusing on the circuit component selection, the modular system may be based on the GaN technology of the semiconductor components. Such a solution is suitable due to the division of the power and voltages to separate individual modules in reduced merit. It also enables us to increase switching frequency several times. Such an approach might reduce the dimensions of used components (magnetic components, capacitors, PCB). Thanks to lower dimensions, it is possible to design converters with smaller PCB, while the volume of a complex modular system would be smaller compared to the non-modular system. Operational parameters of the non-modular system predetermine SiC technology as the main switching component. The switching frequency for these components can be higher compared to standard Si transistors, whereby, considering high voltage and power levels, it is not recommended due to efficiency reduction. Next, Equations (6)–(8) were used for the determination of the values of the main circuit components (Figure 3) affecting the converter volume.

Figure 6 shows the 3D dependency of the values of inductor L and filter capacitor C_{OUT} received using (6)–(8) for the situation when the number of modules and switching frequency vary [17]. At the same time, input/output parameters are relevant for individual module count.

$$L[H] = \frac{V_{in}\,(V_{out} - V_{in})}{\Delta i_L\,I_{out\_max}\,f_{sw}\,V_{out}}$$

(6)

where V_{in} is the input converter voltage (V), V_{out} is the output converter voltage (V), f_{sw} is the switching frequency (Hz), Δi_L is the ripple of inductor current (%), and I_{out_max} is the maximum output current (A).

$$D[\%] = 1 - \frac{V_{in\_min}}{V_{out}} \tag{7}$$

where V_{in_min} is the minimum input converter voltage (V), and V_{out} is the output converter voltage (V).

$$C_{out}[F] = \frac{I_{out\_max}\, D}{\Delta V_{out}\, f_{sw}\, V_{out}} \tag{8}$$

where I_{out_max} is the maximum output current (A), D is the duty cycle (%), f_{sw} is the switching frequency (kHz), ΔV_{out} is the ripple of the output voltage (%), and V_{out} is the output voltage (V).

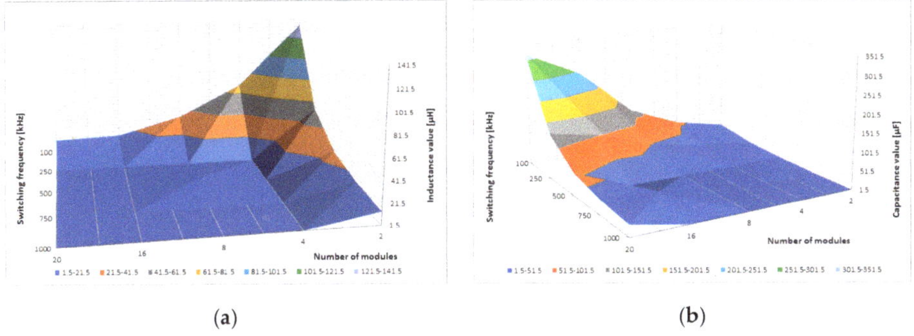

(**a**) (**b**)

Figure 6. Three-dimensional dependency of the value of inductance L (**a**) and the value of capacitance C (**b**) on switching frequency and number of converter modules.

It must be noted that the interpretation considers one module situation. For the whole modular solution, the result must be multiplied by the relevant number of the considered modules.

Table 2 shows input/output parameters that have been included within the calculation of the L and C_{OUT} if real operational conditions are valid. At this point, the need for semiconductor devices is considered for various scenarios. It is seen that for the non-modular solution, a high voltage SiC transistor module is needed. For two and four modules, high-voltage GaN transistors (650 V) must be used, while for a higher number of modules, it is allowed us to utilize 100 V GaN transistors.

Table 2. Table of input/output variations for various bi-BB concepts dependent on Nr. of modules.

Module Count	Output Voltage (V)	Input Voltage (V)	Output Power (W)	V_{DS} (V)	I_D (A)	R_{DSon} (mΩ)
1 (nonmodular)	520	600	10,000	1200 (SiC)	30	75
2	260	300	5000	650 (GaN)	30	55
4	130	150	2500	650 (GaN)	30	55
8	65	75	1250	100 (GaN)	30	15
16	32.5	37.5	625	100 (GaN)	45	15
20	26	30	500.5	100 (GaN)	45	15

At this place, the economic performance, together with efficiency and power density calculation, is given. Initially, Table 3 shows an expert estimation of the investments necessary for the design of proposed solutions of the bi-BB converter. The estimation considers with the whole bill of materials of electronic parts (power semiconductor components, drivers, magnetic components, passive components, and PCB), while the standard distribution network was considered. It is seen that the initial costs of the non-modular DC-DC interleaved converter based on the SiC technology are comparable to the initial costs that are relevant for up to a 16-stage modular DC-DC converter.

Table 3. System costs evaluation for various bi-BB concept dependent on Nr. of modules.

	T	C_{IN}	C_{OUT}	L	PCB	Others	Total
non modular (50 kHz)	20	12	40	150	490	40	712 €
2 modules (100 kHz)	65	11	14	22	320	20	432 €
4 modules (100 kHz)	130	22	26	56	275	22	509 €
8 modules (100 kHz)	83	28	37	80	210	27	438 €
16 modules (100 kHz)	167	38	180	73	320	40	778 €
20 modules (100 kHz)	209	65	210	100	280	50	864 €

Figure 7 shows the graphical interpretation of the so-called qualitative parameters of power semiconductor converters with a dependency on switching frequency. Here, it was defined that these parameters are material costs, efficiency, and expected converter volume (concerning power delivery can be considered as power density). Initially, a non-modular solution is compared that has a dependency on switching frequency. It is seen that with the increase in the switching frequency, the costs together with volume decrease, which is related to the fact that smaller reactive components can be used within the converter's main circuit. Efficiency is almost similar for each of the operating frequencies, as SiC transistors are suitable for the investigated range of this parameter.

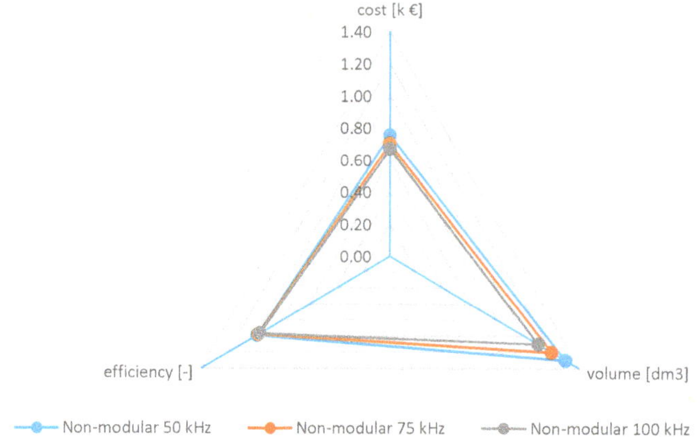

Figure 7. Comparison of qualitative parameters for a non-modular concept in dependency on switching frequency.

Consequently, comparisons are provided between non-modular and modular concepts, while switching frequency is considered as 100 kHz. Considering the volume of the converter (power density), a non-modular solution exhibits performance that is most suitable regarding given switching frequency and input/output parameters that are limited due to power delivery and semiconductor performance (Figure 8). For high power levels, it is expected to operate at lower frequencies in order to prevent unwanted negative impacts (safety reasons, EMC, efficiency reduction, etc.). At the same time, robust semiconductors must be used (IGBT, SiC MOSFETS) [32–34].

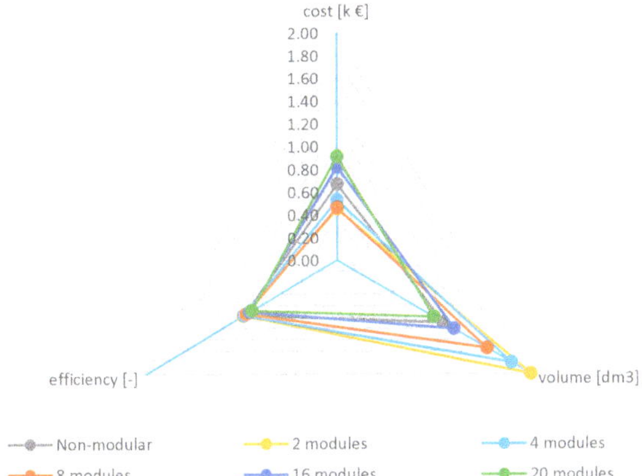

Figure 8. Comparison of qualitative parameters for the non-modular and modular concept for 100 kHz of switching frequency.

The modular solution is not attractive for low switching frequencies due to a power density point of view, which influences the cost of such a solution. On the other side, it is seen that this parameter is best for the case of an eight-module solution. It is related to cheaper power components when the input/output voltage is reduced. Thus, components with lower current/voltage loading can be utilized, and a reasonable number of modules shall be selected (for 20 modules, the cost is very high due to the high number of components). Therefore, the high-switching frequency operation is easy to utilize.

Evaluation of the impact of switching frequency increase is reported in Figures 9 and 10, where only modular solutions are compared for 500 kHz and 1 MHz. With this increase, the volume of the passive components can be visibly reduced. Moreover, when GaN technology is considered, the volume of the semiconductors also minimizes. A GaN-based converter system has a big advantage if a very small volume and weight are required. Typical examples are mobile systems, compact converter systems, or electromobility. From Figures 9 and 10, it is seen that with the increase in switching frequency, the total volume of the modular converter system can be reduced below the volume of the non-modular solution, whereby this is valid from 500 kHz of switching frequency and above four numbers of the modules. The positive impact of frequency increase is the opposite if efficiency is evaluated. For 1 MHz of switching frequency, the efficiency drops below 93% if more than eight modules are used.

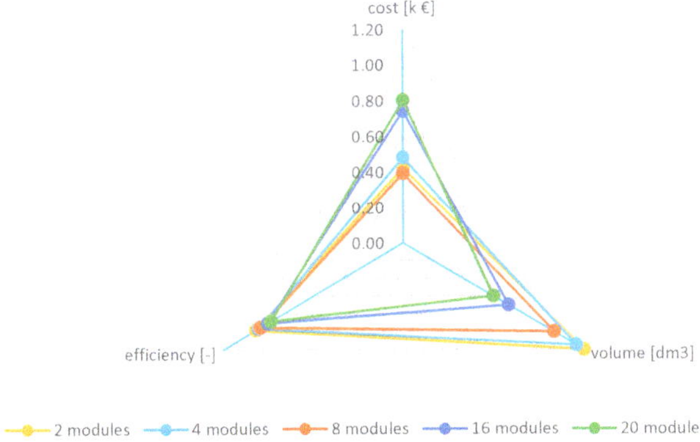

Figure 9. Comparison of qualitative parameters for the modular concept for 500 kHz of switching frequency.

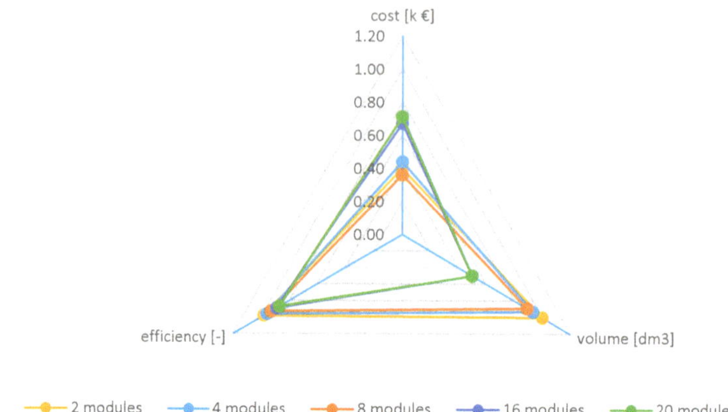

Figure 10. Comparison of qualitative parameters for the modular concept for 1 MHz of switching frequency.

4. Comparison of Operational Properties of Proposed Eight Modules Bi-BB Modular Converter and Bi-BB Non-Modular Interleaved Converter

4.1. Concepts Description

Due to initial validation purposes, the parameters listed in Table 1 were reduced by the power ratio 1:10. Considering similar conditions to the real system, voltage levels were also modified for experimental prototypes of converters (Table 4). The block diagram (Figure 11) indicates the voltage levels selected for practical experiments, while the values are reduced for power delivery of 1 kW full power (the real system operates at 10 kW).

Table 4. Operational parameters of bi-BB converters in a modular and non-modular concept valid for laboratory verifications.

	Input Voltage (Vdc)	Input Current (A)	Output Voltage (Vdc)	Output Current (A)	Switching Frequency (kHz)	Output Power (W)	Phase Shift (°)
Non-modular converter	90–110	10	200	5	150	1000	180
Converter for modular concept	10–14	10	25	5	500	125	45

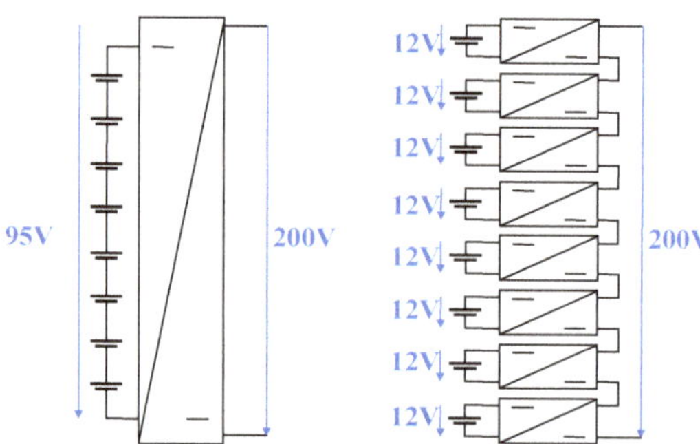

Figure 11. Block diagram of non-modular converter concept (**left**) and modular converter concept (**right**) in reduced power ratio.

The converters in the modular solution are phase-shifted by 360/8° to achieve a low output voltage and current ripple. However, this power ratio emulation is also reflected within component design and selection of the converter's main circuit devices in order to provide us with the most realistic conditions as possible. The non-modular concept utilizes SiC transistors operating at lower switching frequencies (app. 100 kHz) and uses standard inductors. On the other hand, in order to provide an increase in power density performance, the modular concept utilizes low voltage/high-speed GaN transistors (operating over 300 kHz) with planar inductors. This approach shall demonstrate the optimization possibilities of a bidirectional buck-boost converter using a modular converter concept.

The physical prototypes of the converters were designed based on parameters given in Table 4. Table 5 lists the main circuit components used within a non-modular and modular converter prototype.

Table 5. List of used electronic parts for modular and non-modular concept.

	Inductors	Power Transistors	Input Capacitors	Output Capacitors	Gate Drivers
Non-modular converter	2 × PQ40 N87 material, 220uH	Cree C3M0065100K	2 × 150uF/450 V Rubycon + MLCC 100nF	2 × 270uF/450 V Nippon + MLCC 100nF	AD4223 SOIC16
Converter for the modular concept	Bourns 15 μH automotive inductor	GaN systems GS61008T	8 × MLCC 4.7 μF/100V, 2 × Nichicon 1500 μF/35 V electrolytic capacitors	-	LM5113 WSON10

Figure 12 shows a physical sample of proposed bi-BB converters. The non-modular concept uses inductors that are made on PQ40 N87 cores, while the winding is made of isolated copper foil in order to achieve low conduction losses. In order to secure the safe operation of the control system, the isolation on the side of gate drivers was used.

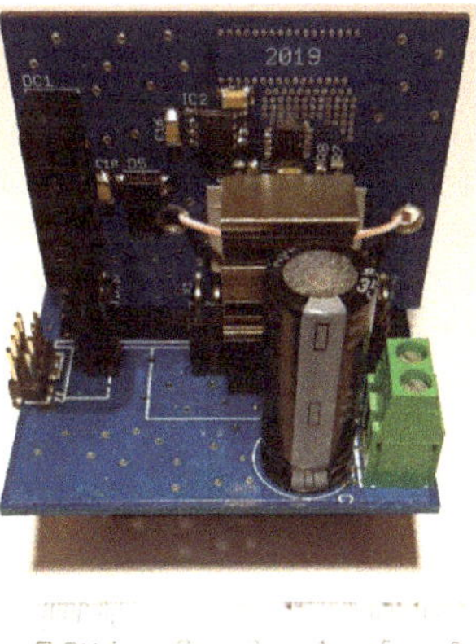

Figure 12. Comparison of cost of non-modular and modular converter concept in dependency on the number of modules and switching frequency.

An experimental prototype of one module that is used for a modular concept, where eight converters with separate inputs and common output are connected to achieve the 200 V of the output voltage, is shown in Figure 12 as well. The proposed module consists of two boards. The horizontal motherboard is composed mostly of filtering components like electrolytic capacitors with MLCC capacitors and power inductors. The vertical board consists of GaN transistors with gate drivers, DC/DC isolated modules, optical isolators, and connector sockets. The vertical board is connected to the motherboard trough socket for better serviceability of measured parameters and more suitable electronic components maintenance.

4.2. Operation Properties Comparisons

The experimental measurements focused on the evaluation of main operational characteristics for both the buck and boost mode of designed bi-BB converter concepts. The evaluations were made separately for efficiency and voltage/current ripples. The laboratory equipment and experimental set-up used within measurements are shown in Figure 13.

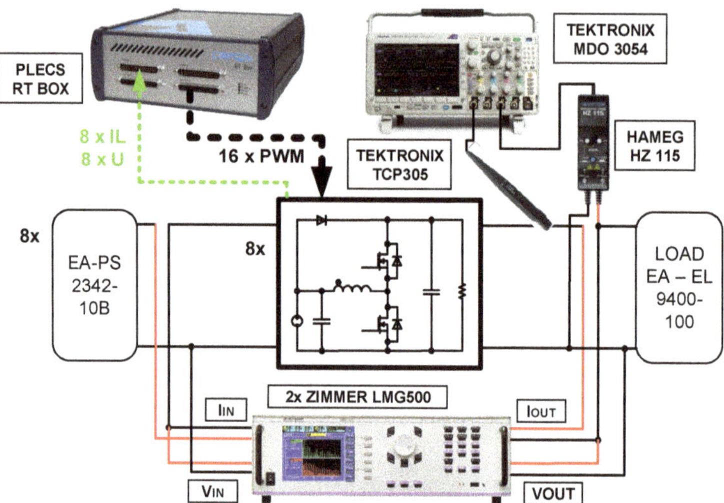

Figure 13. Preview of laboratory equipment and experimental set-up during measurements.

For buck and boost mode, three input voltages were applied, while the investigated variables were analyzed for the whole output power range. Figure 14 shows the efficiency dependency for the boost mode, while the input voltage varied within 90 V and 110 V. Both tested solutions offer almost 97% efficiency, whereby the difference between analyzed converter types is visible in dependency on output power. The input voltage of the modular system is created by a sum of eight voltages on the inputs of individual modules. The efficiency decreases with the increase in output power, while on the other side, the non-modular system has increasing character. These facts are caused due to operational character, for example, due to the three times higher switching frequency of modular concept.

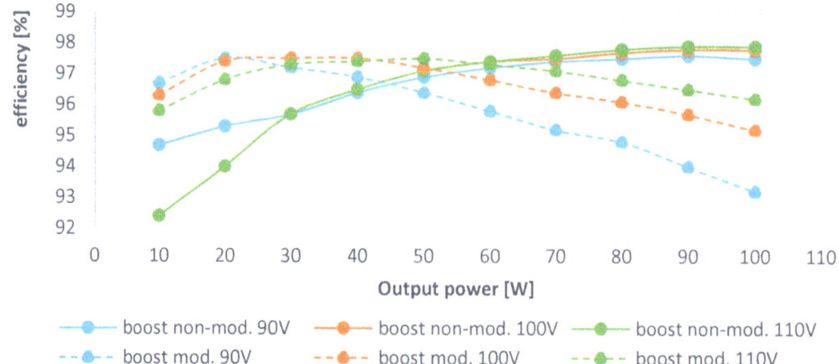

Figure 14. The dependency of efficiency on output power and input voltage for proposed bi-BB converters for boost mode.

Even for the buck mode of operation, the modular system has a decreasing character of efficiency (Figure 15), which is also a cause of the higher number of switching transistors that are used. More transistors cause more hard switching losses and lower efficiency for higher output loads. If both efficiency characteristics for boost and buck mode are analyzed, the modular solution exhibits an advantage below 60% of the nominal power. In contrast, above this point, non-modular solutions become more effective.

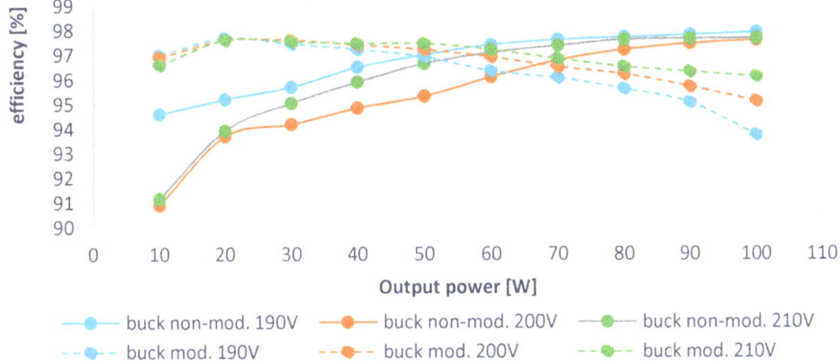

Figure 15. The dependency of efficiency on output power and input voltage for proposed bi-BB converters for buck mode.

Figure 16 shows the output current ripple of the systems in boost mode of operation. The modular system has a current ripple of around 1% and a non-modular system around 3% if nominal power is considered. From the diagram, it is seen that the modular concept has a lower ripple than a non-modular system within the whole power range. This fact is caused by the higher switching frequency in a modular system and interleaved operation given by the 360/8° ratio of control signals.

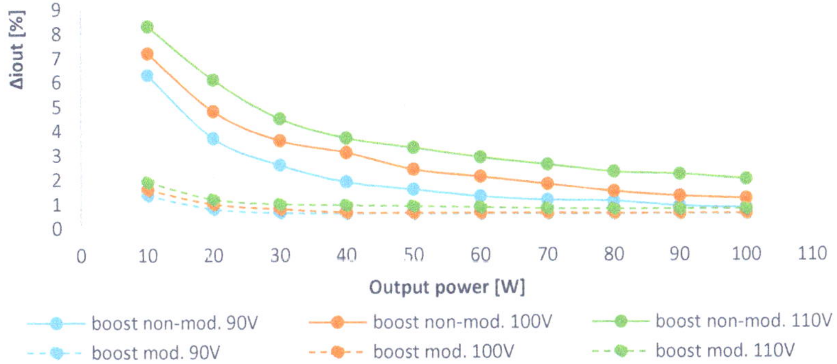

Figure 16. The dependency of output current ripple on output power and input voltage for proposed bi-BB converters for boost mode.

Figure 17 shows the output current ripple of the systems in the buck mode of operation. The modular system again reaches much lower values compared to the non-modular system, while the values of the ripple are below 0.5% if the output power is higher than 30% of the nominal converter's power. During the change in the input voltage, a modular solution exhibits visible independence, while the non-modular system is visibly dependent if the ripple vs. input voltage is analyzed. The lowest ripple for the non-modular solution is achieved at a high output power, which is related to the extension of the duty cycle if the output power is increased.

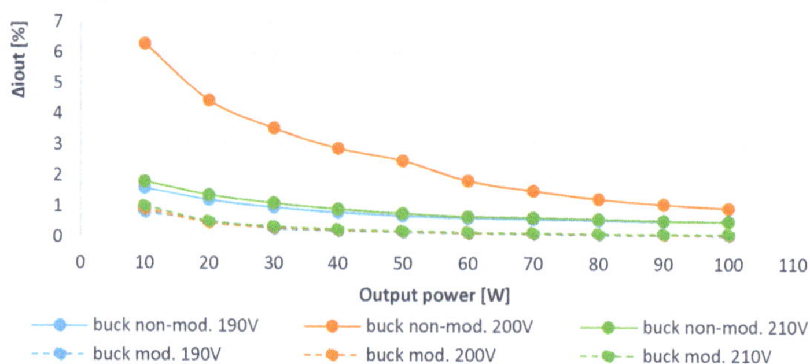

Figure 17. The dependency of output current ripple on output power and input voltage for proposed bi-BB converters for buck mode.

Figures 18 and 19 show the dependency of the output voltage ripple of both concepts in the boost and buck modes of operation. The modular system has a voltage ripple around 0.8% and a non-modular system around 1.6% at the nominal point of operation if buck mode is considered. For boost mode, the modular system has voltage ripple around 1% and the non-modular system around 1.8% at full power. If both operational modes are analyzed (buck and boost), the modular system has a better voltage ripple performance than a non-modular system for any voltage level applied at the input terminals of converters.

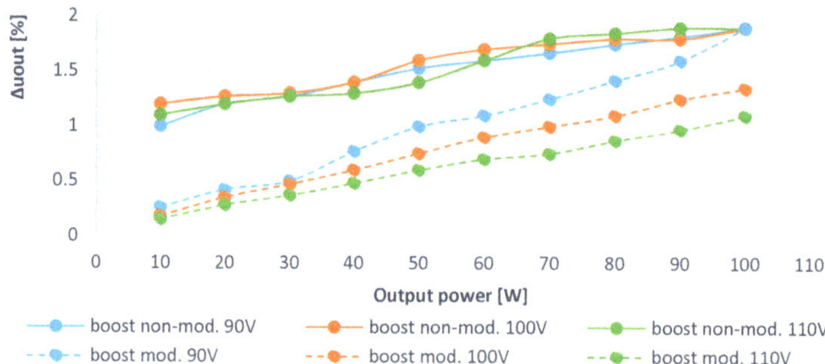

Figure 18. The dependency of output voltage ripple on output power and input voltage for proposed bi-BB converters for boost mode.

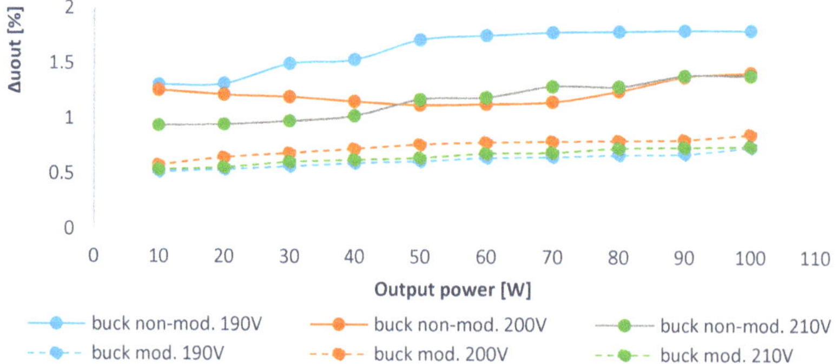

Figure 19. The dependency of output voltage ripple on output power and input voltage for proposed bi-BB converters for buck mode.

Previous analyses showed the advantages and disadvantages of the operational characteristics of designed bi-BB converters. Both have pros and cons related to costs, power density, efficiency performance, as well as the character of electrical variables. Related to the mentioned facts, it is further valuable to investigate the impact of previous research within the target application (i.e., microgrid), where designed converters are used as energy flow control blocks. Seeking higher flexibility for experimental analyses, it is valuable to use hardware in the loop (HIL) simulations, giving more time and space for system optimization.

5. Conclusions

This article deals with the analysis, description of the design, and experimental testing of modular and non-modular bidirectional converter for the energy management block in the energy hub for households. The modular topology was realized by eight modules based on new, very fast GaN transistors technology, which allows increasing the switching frequency up to the range of megahertz. Generally known, this fact causes decreases in the dimensions, volumes, and weight of converters and decreases costs for certain situations. The non-modular topology was also based on new SiC transistors, which also allows the use of high switching frequencies and reduces overall volume and costs. The main electronic parts for designed prototypes, together with specifications of input/output parameters, have been defined for verification of various operational scenarios.

The physical samples of both topologies were successfully tested in laboratory conditions in the full range of output loads, and the efficiency, output voltage ripple, and output current ripple parameters were investigated. The different input voltages were tested for the converters to investigate the behavior of converters for different conditions. The received results and characteristics were discussed in detail within this paper. The modular solution has better voltage and current ripple performance due to the interleaving technique of individual modules. The SiC-based non-modular solution has slightly better efficiency for the full power condition (1000 W). The efficiency characteristics of both topologies are comparable, and the efficiency reaches almost 98%.

Author Contributions: Conceptualization, M.F., methodology, S.K., experimental analysis and set-ups, J.M.; writing—original draft preparation, M.P. All authors have read and agreed to the published version of the manuscript.

Funding: This research was funded by Vega 1/0547/18—Research of possibilities for system optimization of WPT systems. The practical verifications have been provided using infrastructure funded from EU structural funds of the project ITMS 26210120021.

Acknowledgments: The authors would like to thank Slovak national grant agency Vega for the above mentioned financial support as well as for European Structural funds.

Conflicts of Interest: The authors declare no conflict of interest.

Appendix A

The investigation of input current ripple Δi_{INpp} is based on a sum of inductor ripple currents Δi_{L1pp} and Δi_{L2pp}. The input ripples are analyzed separately for the duty ratios: $D \leq 1/2$, $1/2 \leq D < 1$. The necessity for this separation is in the different operation modes of the non-modular converter for $D \leq \frac{1}{2}$ and $1/2 \leq D < 1$ to obtain current ripple value. In the first case, the transistor T1 (lower transistor in a first phase) is on, and the transistor T3 (lower transistor in a second phase) is off. This means that the current i_{L1} in a first phase has a positive slope with a value V_{in}/L (Equation (A1)), and on the other hand, the current i_{L2} in a second phase has a decreasing character with a slope of $(V_{in} - V_{out})/L$, as is shown in Equation (A2). The sum of the ripples Δi_{L1pp} and Δi_{L2pp}, which are ripples of the currents i_{L1} and i_{L2} within the duty ratio period, gives a value of input current ripple Δi_{INpp}. This assumption is valid for the input current of a boost converter.

The same manner can be used for a buck mode within the investigation of output current ripple. It must be stated that the output current of a buck converter is an input current of the non-modular converter as well.

The inductor current ripples and input current ripple for boost interleaved non-modular converter is seen in Figure A1. During the state, as mentioned earlier (T1 is on, and T3 is off), the voltage across the inductor L_1 is equal to V_{in}. From Faraday's law, it is known that the voltage across an inductor is equal to the inductance L times the rate of the current change $V_L = L di/dt$, and therefore for di_{L1} and di_{L2}:

$$di_{L1} = \frac{V_{IN}}{L} dt \tag{A1}$$

$$di_{L2} = \frac{V_{IN} - V_{OUT}}{L} dt \tag{A2}$$

and in this state $dt = DT_S$.

Therefore, the values of ripples for $D < 1/2$ are expressed in Equations (A3)–(A5)

$$\Delta I_{L1pp} = \Delta I_{L1} = \frac{V_{IN}}{L} DT_S \tag{A3}$$

$$\Delta I_{L2pp} = \Delta I_{L2} = \frac{V_{IN} - V_{OUT}}{L} DT_S \tag{A4}$$

Then, the solution for input ripple current is as follows:

$$\Delta I_{INpp} = \Delta I_{L1pp} + \Delta I_{L2pp} = \frac{V_{OUT}}{L} DT_S (1 - 2D) \tag{A5}$$

The solution of input current ripple for the duty ratio within a range of $1/2 \leq D < 1$ is shown in Figure A2.

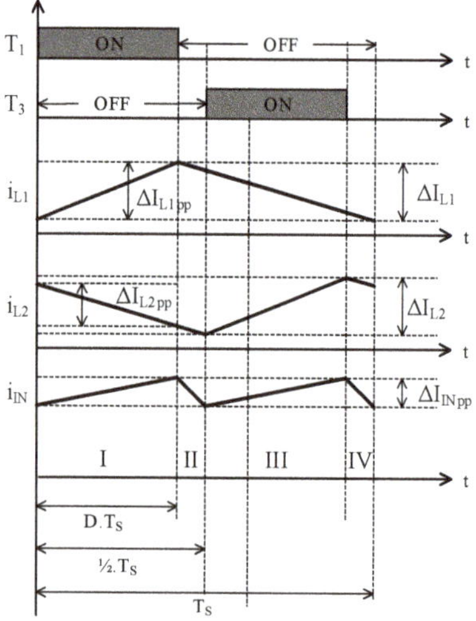

Figure A1. Input and inductor currents for $D < \frac{1}{2}$.

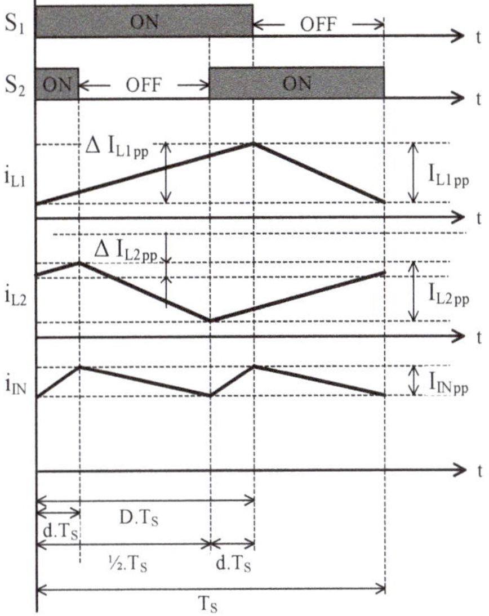

Figure A2. Input and inductor currents for $1/2 \leq D < 1$.

The inductor current ripples are given in Equations (A6) and (A7). The procedure for obtaining Equations (A6) and (A7) is the same as in the previous case. The difference is that the $dt = dT_S$. The new parameter d is involved in the calculation because the input current ripple occurs within the interval dT_S. The parameter d is expressed in Equation (A8). The procedure for obtaining an equation is

$$\Delta I_{L1pp} = \Delta I_{L1} = \frac{V_{IN}}{L} dT_S \tag{A6}$$

$$\Delta I_{L2pp} = \Delta I_{L2} = \frac{V_{IN}}{L} dT_S \tag{A7}$$

$$d = D - \frac{1}{2} \tag{A8}$$

Then, the solution for the input current ripple is a sum of the inductor current ripples, Equation (A9).

$$\Delta I_{INpp} = \Delta I_{L1pp} + \Delta I_{L2pp} = \frac{V_{OUT}}{L} DT_S \left(D - \frac{1}{2}\right)(2 - 2D) \tag{A9}$$

These solutions for $D \leq \frac{1}{2}$ and $1/2 \leq D < 1$ are also shown in Table A1. The number of phases is two, and interval I and interval II are considered. It must be stated that with an increase in the number of phases, the number of intervals also increases. This is due to the greater number of operating modes of the converter. Therefore, the n-phase converter is divided into n intervals.

The same assumption is valid for the converter in a buck mode. The difference is only in output V_{out} and input voltage V_{in}. It should be noted that the output voltage of the boost converter is the input voltage of the buck converter. Therefore, for a non-modular converter, the equations are the same. Then, the input current ripples for the two-, three-, four- and n-phase non-modular converters in a boost and buck mode are shown in a Tables A1 and A2, respectively.

Table A1. The equations of output current ripples for the 2-,3-, 4- and n-phase non-modular boost converter.

Number of Phases	Interval I	Interval II	Interval III	Interval IV	Interval m
1					
2	$\Delta I_{IN} = \frac{V_{OUT}}{L} DT_S(1-2D)$	$\Delta I_{IN} = \frac{V_{OUT}}{L} T_S\left(D - \frac{1}{2}\right)(2-2D)$			
3	$\Delta I_{IN} = \frac{V_{OUT}}{L} DT_S(1-3D)$	$\Delta I_{IN} = \frac{V_{OUT}T_S}{L}\left(D - \frac{1}{3}\right)(2-3D)$	$\Delta I_{IN} = \frac{V_{OUT}T_S}{L}\left(D - \frac{2}{3}\right)(3-3D)$		
4	$\Delta I_{IN} = \frac{V_{OUT}T_S}{L} D(1-4D)$	$\Delta I_{IN} = \frac{V_{OUT}T_S}{L}\left(D - \frac{1}{4}\right)(2-4D)$	$\Delta I_{IN} = \frac{V_{OUT}T_S}{L}\left(D - \frac{2}{4}\right)(3-4D)$	$\Delta I_{IN} = \frac{V_{OUT}T_S}{L}\left(D - \frac{3}{4}\right)(4-4D)$	
...	...	...	...	...	
n	$\Delta I_{IN} = \frac{V_{OUT}T_S}{L} D(1-nD)$	$\Delta I_{IN} = \frac{V_{OUT}T_S}{L}\left(D - \frac{1}{n}\right)(2-nD)$	$\Delta I_{IN} = \frac{V_{OUT}T_S}{L}\left(D - \frac{2}{n}\right)(3-nD)$	$\Delta I_{IN} = \frac{V_{OUT}T_S}{L}\left(D - \frac{3}{n}\right)(4-nD)$	$\Delta I_{IN} = \frac{V_{OUT}T_S}{L}\left(D - \frac{m}{n}\right)(m-nD)$

Table A2. The equations of output current ripples for the 2-,3-, 4- and n-phase.

N.	Interval I	Interval II	Interval III	Interval IV	Interval m
1					
2	$\Delta I_{OUT} = \frac{V_{IN}}{L} DT_S(1-2D)$	$\Delta I_{OUT} = \frac{V_{IN}}{L} T_S\left(D - \frac{1}{2}\right)(2-2D)$			
3	$\Delta I_{OUT} = \frac{V_{IN}}{L} DT_S(1-3D)$	$\Delta I_{OUT} = \frac{V_{IN}T_S}{L}\left(D - \frac{1}{3}\right)(2-3D)$	$\Delta I_{OUT} = \frac{V_{IN}T_S}{L}\left(D - \frac{2}{3}\right)(3-3D)$		
4	$\Delta I_{OUT} = \frac{V_{IN}T_S}{L} D(1-4D)$	$\Delta I_{OUT} = \frac{V_{IN}T_S}{L}\left(D - \frac{1}{4}\right)(2-4D)$	$\Delta I_{OUT} = \frac{V_{IN}T_S}{L}\left(D - \frac{2}{4}\right)(3-4D)$	$\Delta I_{OUT} = \frac{V_{IN}T_S}{L}\left(D - \frac{3}{4}\right)(4-4D)$	
...	...	...	...	...	
n	$\Delta I_{OUT} = \frac{V_{IN}T_S}{L} D(1-nD)$	$\Delta I_{OUT} = \frac{V_{IN}T_S}{L}\left(D - \frac{1}{n}\right)(2-nD)$	$\Delta I_{OUT} = \frac{V_{IN}T_S}{L}\left(D - \frac{2}{n}\right)(3-nD)$	$\Delta I_{OUT} = \frac{V_{IN}T_S}{L}\left(D - \frac{3}{n}\right)(4-nD)$	$\Delta I_{OUT} = \frac{V_{IN}T_S}{L}\left(D - \frac{m}{n}\right)(m-nD)$

Appendix B

The inductor and output capacitor currents i_{L1}, i_{L2}, i_{C1}, and i_{C2} are depicted in Figure A3 for two modules of the modular converter. It is seen that the inductor current ripple ΔI_{L1} in one-phase is equal to capacitor current ripple ΔI_{Coff} during the period that the transistor of the relevant phase is switched off. Therefore, according to the previous procedure, the following equation is valid:

$$\Delta I_{L1} = \Delta I_{Coff} = \frac{V_{IN} - V_{OUT}}{Lf_S}(1 - D) \tag{A10}$$

In the modular converter, the topology simplification can be used. The output capacitor is connected in series; then, the final value of the output capacitor is eight times lower. If we consider one output capacitor, the waveform of the output capacitor current is displayed in Figure A3 with a blue line. It is seen that the modified period of the output capacitor ripple current is one eighth of the switching period. This is due to the equal phase-shifting of the eight-module converter. Then, the output capacitor current ripple ΔI_{Cout} is dependent on a slope of the inductor/capacitor current and the modified period.

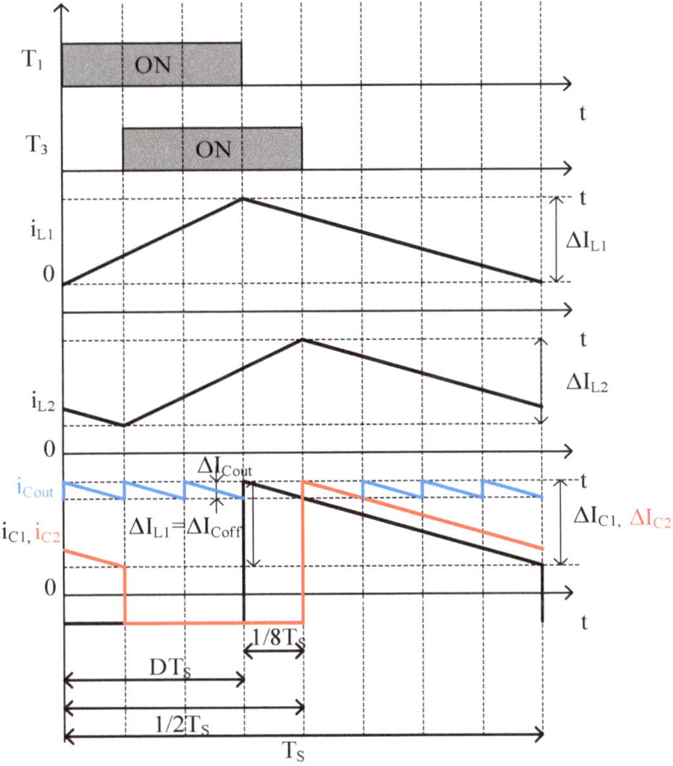

Figure A3. Output capacitor and inductor current of modular converter.

The duration of the slope of the inductor current is $(1 - D)$ TS, and the ripple is as follows.

$$\Delta I_L = \frac{V_{out} - V_{in}}{Lf_{sw}}(1 - D) \tag{A11}$$

The duration of the output capacitor current is $1/8\ T_S$. Then we can write:

$$\Delta I_C = \frac{1}{8}\frac{V_{out} - V_{in}}{L f_{sw}} \tag{A12}$$

References

1. Forouzesh, M.; Siwakoti, Y.P.; Gorji, S.A.; Blaabjerg, F.; Lehman, B. Step-up DC-DC converters: A comprehensive review of voltage-boosting techniques, topologies, and applications. *IEEE Trans. Power Electron.* **2017**, *32*, 9143–9178. [CrossRef]
2. Dawidziuk, J. Review and comparison of high-efficiency high power boost DC/DC converters for photovoltaic applications. *Bull. Pol. Acad. Sci. Tech. Sci.* **2011**, *59*, 499–506. [CrossRef]
3. Davison, M.J.; Summers, T.J.; Townsend, C.D. A review of the distributed generation landscape, key limitations of traditional microgrid concept & possible solution using an enhanced microgrid architecture. In Proceedings of the 2017 IEEE Southern Power Electronics Conference (SPEC), Puerto Varas, Chile, 4–7 December 2017; pp. 1–6. [CrossRef]
4. Abdelgawad, H.; Sood, V.K. A Comprehensive Review on Microgrid Architectures for Distributed Generation. In Proceedings of the 2019 IEEE Electrical Power and Energy Conference (EPEC), Montreal, QC, Canada, 16–18 October 2019; pp. 1–8. [CrossRef]
5. Gupta, P.; Ansari, M.A. Analysis and Control of AC and Hybrid AC-DC Microgrid: A Review. In Proceedings of the 2019 2nd International Conference on Power Energy, Environment and Intelligent Control (PEEIC), Greater Noida, India, 18–19 October 2019; pp. 281–286. [CrossRef]
6. Brahmendra Kumar, G.V.; Palanisamy, K. A Review on Microgrids with Distributed Energy Resources. In Proceedings of the 2019 Innovations in Power and Advanced Computing Technologies (i-PACT), Vellore, India, 22–23 March 2019; pp. 1–6. [CrossRef]
7. Kondrath, N. Bidirectional DC-DC converter topologies and control strategies for interfacing energy storage systems in microgrids: An overview. In Proceedings of the 2017 IEEE International Conference on Smart Energy Grid Engineering (SEGE), Oshawa, ON, Canada, 14–17 August 2017; pp. 341–345. [CrossRef]
8. Lai, C.M.; Lin, Y.C.; Lee, D. Study and implementation of a two-phase interleaved bidirectional dc/dc converter for vehicle and DC-microgrid systems. *Energies* **2015**, *8*, 9969–9991. [CrossRef]
9. Esashika, H.; Natori, K.; Sato, Y. A Universal Control Method to Realize Plug-and-Play Converters for Microgrids. In Proceedings of the 2019 10th International Conference on Power Electronics and ECCE Asia (ICPE 2019—ECCE Asia), Busan, Korea, 27–30 May 2019; pp. 1–7.
10. Ovcarcik, R.; Spanik, P.; Pavlanin, R. DC/DC converters used for a high input voltage based on a half-bridge topology. In Proceedings of the 6th European Conference of TRANSCOM: 6th European Conference of Young Research and Science Workers in Transport and Telecommunications, Žilina, Slovakia, 27–29 June 2005; pp. 57–62.
11. Radika, P. High-efficiency DC-DC boost converter with passive regenerative snubber. *J. Electr. Eng. Technol.* **2014**, *9*, 501–507. [CrossRef]
12. Perdulak, J.; Kovac, D.; Kovacova, I.; Ocilka, M.; Gladyr, A.; Mamchur, D.; Zachepa, I.; Vince, T.; Molnar, J. Effective utilization of photovoltaic energy using multiphase boost converter in compare with single phase boost converter. *Commun. Sci. Lett. Univ. Zilina* **2013**, *15*, 32–38.
13. Kim, D.Y.; Won, I.K.; Lee, J.H.; Won, C.Y. Efficiency Improvement of Synchronous Boost Converter with Dead Time Control for Fuel Cell-Battery Hybrid System. *J. Electr. Eng. Technol.* **2017**, *12*, 1891–1901.
14. De Caro, S.; Testa, A.; Triolo, D.; Cacciato, M.; Consoli, A. Low input current ripple converters for fuel cell power units. In Proceedings of the 2005 European Conference on Power Electronics and Applications, Dresden, Germany, 11–14 September 2005; pp. 1–10.
15. Choe, G.Y.; Kim, J.S.; Kang, H.S.; Lee, B.K. An optimal design methodology of an interleaved boost converter for fuel cell applications. *J. Electr. Eng. Technol.* **2010**, *5*, 319–328. [CrossRef]
16. Cacciato, M.; Consoli, A.; Attanasio, R.; Gennaro, F. Multi-stage converter for domestic generation systems based on fuel cells. In Proceedings of the 41st IAS Annual Meeting (IEEE Industry Applications Society), Tampa, FL, USA, 8–12 October 2006; pp. 230–235.

17. Kascak, S. Analysis of Bidirectional converter with coupled inductor for electric drive application. In Proceedings of the 2016 International Conference on Mechatronics, Control and Automation Engineering, Heilongjiang, China, 7–10 August 2016; Atlantis Press: Paris, France, 2016; pp. 229–232.

18. Cacciato, M.; Caricchi, F.; Giuhlii, F.; Santini, E. A critical evaluation and design of bi-directional DC/DC converters for super-capacitors interfacing in fuel cell applications. In Proceedings of the Conference Record of the 2004 IEEE Industry Applications Conference, 2004. 39th IAS Annual Meeting, Seattle, WA, USA, 3–7 October 2004; Volume 2, pp. 1127–1133. [CrossRef]

19. Sallan, J.; Villa, J.L.; Llombart, A.; Sanz, J.F. Optimal Design of ICPT Systems Applied to Electric Vehicle Battery Charge. *IEEE Trans. Ind. Electron.* **2009**, *56*, 2140–2149. [CrossRef]

20. Kozacek, B.; Kostal, J.; Frivaldsky, M. Analysis of Figure Of Merit—Power transistor's qualitative parameter. In Proceedings of the 2015 16th International Scientific Conference on Electric Power Engineering (EPE), Kouty nad Desnou, Czech Republic, 20–22 May 2015; pp. 718–722. [CrossRef]

21. Frivaldsky, M.; Drgona, P.; Kozacek, B.; Piri, M.; Pridala, M. Critical component's figure of merite influence on power supply unit efficiency. In Proceedings of the 2016 ELEKTRO, Strbske Pleso, Slovakia, 16–18 May 2016; pp. 147–151. [CrossRef]

22. Cacciato, M.; Consoli, A.; Crisafulli, V.; Abbate, N.; Vitale, G. Digital controlled bidirectional DC/DC converter for electrical and hybrid vehicles. In Proceedings of the 14th International Power Electronics and Motion Control Conference EPE-PEMC 2010, Ohrid, Macedonia, 6–8 September 2010; pp. 111–116. [CrossRef]

23. Zaskalicky, P. Complex fourier series analysis of a three-phase induction motor supplied by a three-phase inverter with PWM output voltage control. In Proceedings of the 2014 16th International Power Electronics and Motion Control Conference and Exposition, Antalya, Turkey, 21–24 September 2014; pp. 1183–1188. [CrossRef]

24. Cacciato, M.; Consoli, A.; Crisafulli, V.; Vitale, G.; Abbate, N. A new resonant active clamping technique for bi-directional converters in HEVs. In Proceedings of the 2010 IEEE Energy Conversion Congress and Exposition, Atlanta, GA, USA, 12–16 September 2010; pp. 1436–1441.

25. Chakraborty, S.; Wu, H.N.; Hasan, M.M.; Tran, D.D.; El Baghdadi, M.; Hegazy, O. DC-DC Converter Topologies for Electric Vehicles, Plug-in Hybrid Electric Vehicles and Fast Charging Stations: State of the Art and Future Trends. *Energies* **2019**, *12*, 1569. [CrossRef]

26. Li, W.H.; He, X.N. Review of Non-isolated High-Step-Up DC/DC Converters in Photovoltaic Grid-Connected Applications. *IEEE Trans. Ind. Electron.* **2011**, *58*, 1239–1250. [CrossRef]

27. Garrigos, A.; Sobrino-Manzanares, F. Interleaved multi-phase and multi-switch boost converter for fuel cell application. *Int. J. Hydrog. Energy* **2015**, *40*, 8419–8432. [CrossRef]

28. Gu, Y.; Zhang, D.L. Interleaved boost converter with ripple cancellation network. *IEEE Trans. Power Electron.* **2013**, *28*, 3860–3869. [CrossRef]

29. Kascak, S.; Prazenica, M.; Jarabicova, M.; Konarik, R. Four/phase interleaved boost converter analysis and verification. *Acta Electrotech. Inform.* **2018**, *18*, 35–40. [CrossRef]

30. Lin, P.; Zhao, T.; Wang, B.; Wang, Y.; Wang, P. A Semi-Consensus Strategy Toward Multi-Functional Hybrid Energy Storage System in DC Microgrids. *IEEE Trans. Energy Convers.* **2020**, *35*, 336–346. [CrossRef]

31. Wang, B.; Wang, Y.; Xu, Y.; Zhang, X.; Gooi, H.B.; Ukil, A.; Tan, X. Consensus-based Control of Hybrid Energy Storage System with a Cascaded Multiport Converter in DC Microgrids. *IEEE Trans. Sustain. Energy* **2019**. [CrossRef]

32. Yoo, H.D.; Markevich, E.; Salta, G.; Sharon, D.; Aurbach, D. On the challenge of developing advanced technologies for electrochemical energy storage and conversion. *Mater. Today* **2014**, *17*, 110–121. [CrossRef]

33. Chiu, H.J.; Lin, L.W.; Pan, P.L.; Tseng, M.H. A novel rapid charger for lead-acid batteries with energy recovery. *IEEE Trans. Power Electron.* **2006**, *21*, 640–647. [CrossRef]

34. Dudrik, J.; Pástor, M.; Lacko, M.; Žatkovič, R. Zero-Voltage and Zero-Current Switching PWM DC–DC Converter Using Controlled Secondary Rectifier with One Active Switch and Nondissipative Turn-Off Snubber. *IEEE Trans. Power Electron.* **2018**, *33*, 6012–6023. [CrossRef]

Article

SiC-MOSFET and Si-IGBT-Based dc-dc Interleaved Converters for EV Chargers: Approach for Efficiency Comparison with Minimum Switching Losses Based on Complete Parasitic Modeling

Jelena Loncarski [1,2,*,†], Vito Giuseppe Monopoli [1,†] and Giuseppe Leonardo Cascella [1,†] and Francesco Cupertino [1,†]

[1]   Department of Electrical and Information Engineering, Politecnico di Bari, Via E. Orabona, 4, 70125 Bari, Italy; vitogiuseppe.monopoli@poliba.it (V.G.M.); giuseppeleonardo.cascella@poliba.it (G.L.C.); francesco.cupertino@poliba.it (F.C.)

[2]   Department of Engineering Sciences, Division of Electricity, Uppsala University, S-751 21 Uppsala, Sweden

*   Correspondence: jelena.loncarski@poliba.it

†   These authors contributed equally to this work.

Received: 5 August 2020; Accepted: 27 August 2020; Published: 3 September 2020

Abstract: Widespread dissemination of electric mobility is highly dependent on the power converters, storage systems and renewable energy sources. The efficiency and reliability, combined with the emerging and innovative technologies, are crucial when speaking of power converters. In this paper the interleaved dc–dc topology has been considered for EV charging, due to its improved reliability. The efficiency comparison of the SiC-MOSFET and Si-IGBT-based converters has been done on wide range of switching frequency and output inductances. The interleaved converters were considered with the optimal switching parameters resulting from the analysis done on a detailed parasitic circuit model, ensuring minimum losses and maintaining the safe operating area. The analysis included the comparison of different inductors, and for the selected ones the complete system efficiency and cost were conducted. The results indicate the benefits when SiC-MOSFETs are applied to the interleaved dc–dc topology for wide ranges of output inductances and switching frequencies, and most importantly, they offer lower total volume but also total cost. The realistic and dynamic models of power devices obtained from the manufacturer's experimental tests have been considered in both LTspice and PLECS simulation tools.

Keywords: electric vehicle (EV); fast charging; interleaved dc–dc converter, SiC devices; Si devices

---

1. Introduction

Environmental concerns and green energy goals have led to the growing interest in electric vehicle (EV) and plug-in hybrid electric vehicle (PHEV) technologies, since they offer reduced greenhouse emissions [1]. Governments worldwide are encouraging electric transportation with public policies and new standards. However, the critical points in the large changeover from combustion engine to EVs are still the batteries, suitable chargers and charging infrastructure [2].

In general, the batteries can be charged by on-board chargers (slow chargers) and off-board ones (fast chargers), both unidirectionally or bidirectionally. In either case, both the topologies and power ratings of the converters differ greatly [3]. EV battery chargers include both a dc–dc stage (typically buck/boost or switch-mode converters) and an ac–dc stage (controlled or uncontrolled rectifiers). The ac–dc stage should ensure high power quality operation, while the dc–dc stage should ensure high controllability and operability with constant current and voltage on battery side [4]. The on-board chargers can be conductive (with direct contact) or inductive (transfer power magnetically) [5]. These chargers are nowadays widely employed, but are constrained by weight, space and cost. Nonetheless, all battery chargers have to be of high efficiency and reliability, and high power density. Their operation depends on components, sizing, control and switching strategies.

Fast charging refers to the off-board charging stations with at least 50 kW rated power [6]. Off-board chargers are less subjects of restrictions when compared to on-board chargers and feature high charging rates [7]. Various solutions for both ac–dc and dc–dc stages have been proposed and proven promising [8–12]. The concept which can be viably adopted is interleaved (also multiphase) converters, leading to higher efficiency, with possibilities of current ripple cancellation, and reduction in EMI and output filter size. [8,11,12]. Using interleaving, the power stage of a converter is symmetrically distributed into several smaller power stages. Consequently, the sizes of the passive components are reduced which allows a reduction in the overall cost of the system. Moreover, the system reliability can also be improved, due to the modularity that interleaved converters possess. Many interleaved topologies have been studied, both in ac–dc and dc–dc converter applications [13–15], and for EV charging [11,12,16,17]. When it comes to dc–dc converter applications, in [15] the multiphase interleaved converter for high current applications has been presented, operating in a wide load range. It has been proven that the phase-shift interleaving operation brings the decrease in the current ripple in the multiphase inverter stage. A three-phase interleaved dc–dc converter for EV fast charging with an effective control strategy has been proposed in [11,12], providing ripple-free output current in a wide output voltage range. In [16] the interleaving concept has been used to keep the input current ripple to a very low level in current-fed converter, and can viably be used in PHEV applications.

Recently, the wide bandgap semiconductor devices, such as silicon carbide (SiC), have become increasingly popular due to many benefits they offer compared to their silicon (Si) counterparts: lower switching losses and consequently high-switching capabilities, low on resistance and increased junction temperature [18]. With this in mind, and considering the benefits that an interleaved dc–dc converter offers, it is crucial to conduct the fair comparison of this specific topology with SiC metal-oxide-semiconductor field-effect transistor (SiC-MOSFET) and Si insulated-gate bipolar transistor (Si-IGBT) devices. Even though there have been many comparisons conduced in the literature for other topologies [19–21], for this particular topology not many studies are available. This kind of analysis on interleaved converter was firstly introduced in [22], where the efficiencies of SiC-MOSFET and Si-IGBT-based interleaved converters were compared, but considering only one value of the output inductance and switching frequency. This paper extends this analysis and considers the efficiency comparison on a wide range of output inductances and switching frequencies for both SiC-MOSFET and Si-IGBT-based converters. The interleaved topology was considered for the dc–dc power stage, having the possibility of the voltage and current ripple reduction in the battery-side and making it relevant for EV chargers. The comparison with different devices (SiC-MOSFET and Si-IGBT) is crucial in order to ease the decision process of power converter designers, and choose the right topology for the specific application. Moreover, the analysis includes the comparison of the total system losses and costs, including both converters and various output inductors. The goal was to have the maximum efficiency and minimum losses for both converters, maintaining the safe operating area of each device. Consequently, the most optimal switching condition has been derived from the complete parasitic model, based on experimental measurements. Namely, the minimum values of the external gate resistances have been determined by implementing the parasitic model in LTpice (Version XVII, Analog Devices, Norwood, MA, USA) simulation tool. Next, these resistances were used with double pulse testing in LTspice simulation tool on realistic and dynamic spice models of the respective power switches [23], from which the respective conduction and switching losses have been obtained for each device. These losses were finally implemented in complete 3-leg converter simulation in PLECS (Version 4.2.6, Plexim, Zurich, CH) , from which the loss comparison was conducted, with access to the steady state being facilitated.

2. Dc–dc Interleaved Converter

In Figure 1a is given the fast charger topology for electric vehicles. It consists of a three-phase ac–dc converter connected to an interleaved dc–dc converter with three legs having a three-phase structure. This structure was firstly introduced in [11,12]. The interleaved dc–dc converter is made by basic elements which are paralleled. It can be seen as a well-known two-level, three-phase configuration with the output inductor L_{out} and input dc-link capacitor C_{dc}. By having a rather simple structure, this converter features high reliability. The scalability is ensured as well, due to its modular structure.

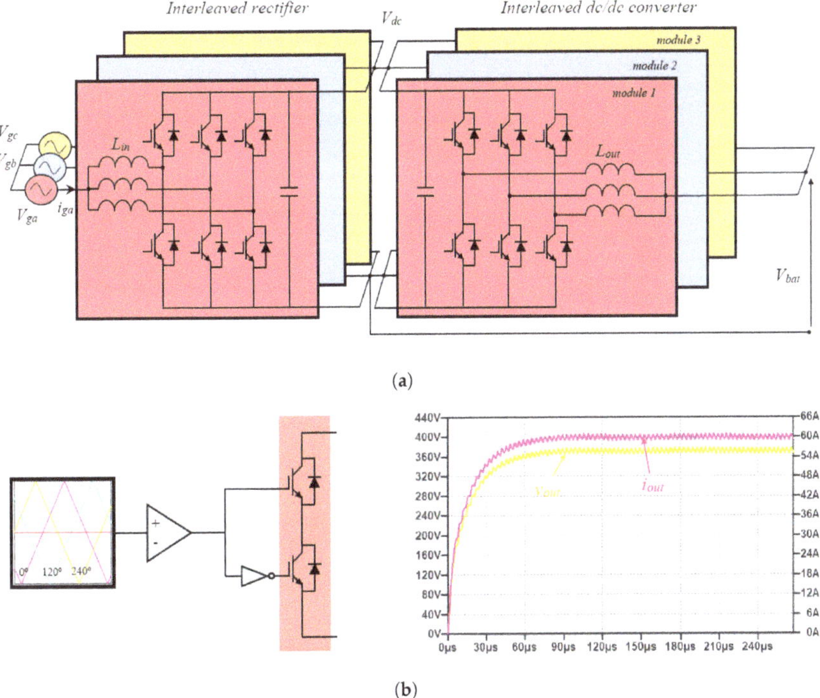

Figure 1. Fast charger topology for EVs: (**a**) three-phase ac–dc converter and an interleaved dc–dc converter, and (**b**) control signals and typical output waveforms.

The ac–dc converter is connected to the grid and has the task of regulating the dc-link voltage. The interleaved dc–dc converter is able to provide the desired output current by controlling the output voltage. The converter's output current is equally shared among the legs. It is possible to achieve the output current ripple minimization with the control strategy [11]. Namely, with the carrier phase-shift of 360° divided by the number of legs and proper control, it is possible to achieve the minimum of the output current ripple, as shown in Figure 1b in the case of 3-leg converter. In Figure 1b the typical output waveforms of the interleaved converter are also shown.

3. Parasitic Model and Safe Operating Area

The interleaved converter permits the use of standard and reliable three-phase power switch modules. The sizes of the ac–dc stage and dc–dc stage are similar, which can be tracked back to sizing only a two-switch leg, which makes the design process rather simple. The simple two-switch leg can be scaled up by adding more legs. In this way the cost and the complexity of the system can be reduced.

In order to have the maximum efficiency in terms of fastest switching and minimum switching losses of the two converters (one with Si-IGBT and other with SiC-MOSFET devices), it is important to carefully select the external gate resistance R_{g_ext}, since it can greatly influence the switching losses. The analysis in this regard started from the realistic and complete parasitic model of the three-leg converter, relying on the design practice and experimental measurements done on a existing printed circuit board (PCB) with connection cables for one converter leg. In Figure 2 is presented the three-leg model of the interleaved converter, taking into account the dc power supply V_{dc} with the cable connection for the PCB (in blue); the dc-link capacitors C_{dclink} (details on the dc-link ralization and capacitor parasitics in yellow) with the parasitic inductance of the capacitor bank PCB trace; the parasitic inductances of the PCB traces connecting the different legs and switches; and the impedance of the cable connecting the output inductor L_{out}. In Table 1 are given the values that have been obtained on the basis of the experimental measurements for the dc power supply and load connection cable, and calculated for the PCB traces with the standard trace thickness 105 µm and width of 0.93 cm required for the considered currents [24]:

$$\frac{L}{l} = 2 \left(ln \left(\frac{2l}{w+t} \right) + \frac{1}{2} \right) \ (\text{nH/cm}) \tag{1}$$

where L is the trace inductance, l is the trace length, w is the width and t is the thickness of the PCB trace. The considered values of PCB traces were carefully selected and are based on the real power electronic circuit design.

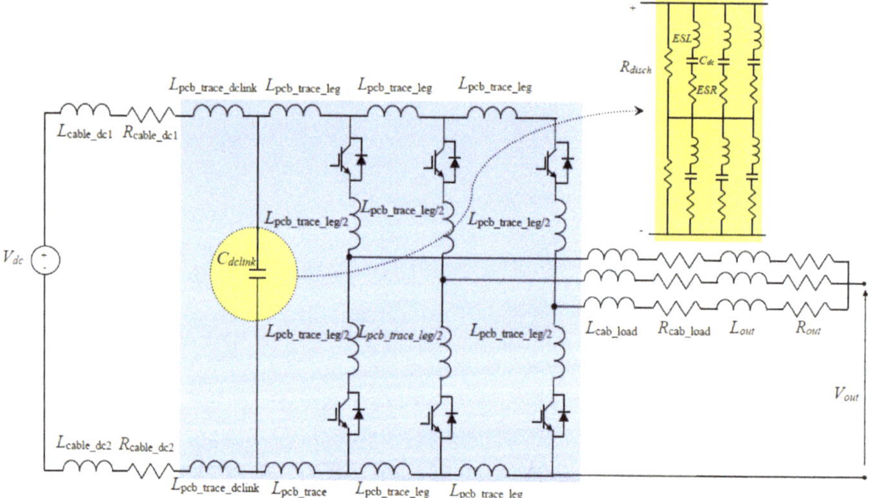

Figure 2. Dc–dc interleaved converter: full parasitic model of 3 legs.

Table 1. PCB parameters.

Parameter (Unit)	Value
L_{cable_dc1} (µH)	1.77
R_{cable_dc1} (mΩ)	8.92
L_{cable_dc2} (µH)	1.74
R_{cable_dc2} (mΩ)	8.79
$L_{pcb\_trace\_dclink}$ (nH)	36.5
$L_{pcb\_trace\_leg}$ (nH)	66.5
$L_{pcb\_trace\_leg/2}$ (nH)	33.25
L_{cab_load} (µH)	0.834
R_{cab_load} (mΩ)	4.21
Thin film capacitor C_{dc} (µH)	15
ESR (mΩ)	4.8
ESL (nH)	12.4
R_{disch} (kΩ)	100

The circuit has been implemented in LTspice simulation tool. The real models of the powers switches were used for this analysis, i.e., the 1200 V SiC device SCT3080KLHR Rohm [25] with the SiC Schottky anti-parallel diode SCS220KG Rohm [26] for the SiC-MOSFET-based converter, and 1200 V Si-IGBT device RGS50TSX2DHR Rohm [27] for the IGBT-based converter (model with fast recovery diode), having the main parameters listed in Table 2, giving quick access to the sizing of different devices. The complete model of three-leg converter has been considered, as shown in Figure 2. The analysis included a set of switching frequencies and output inductors, as listed in Table 3. The goal was to obtain the minimum external gate resistance value for each set of L_{out} and f_{sw}, in order to enable fastest switching with minimum losses, but staying within the safe operating area (SOA) of the device. What has also been considered is the internal resistance of the gate driver BM6105AFW-LBZ (Rohm) R_{int} of 1.5 Ω, which can be used for both SiC-MOSFET and Si-IGBT devices. With this in mind, the gate-source voltage of the SiC-MOSFET-based converter was $-4/18$ V ensuring the optimal turn on and turn off of the device, while for the Si-IGBT inverter the gate-emitter voltage 0/15 V has been used.

The resulting required minimum gate resistances to be added externally to the gate driver for SiC-MOSFET and Si-IGBT-based converters are shown in Table 3, together with the simulation parameters that are used for LTspice simulation such as the values of the $L_{out} - f_{sw}$ pairs. The minimum required values of the gate resistances have been determined in the LTspice simulation by increasing the value of the gate resistance (starting from 1.5 Ω), and taking the first value for which the SOA region is satisfied. The SiC-MOSFET-based converter requires no additional gate resistance (apart from the internal resistance of the gate driver), while the Si-IGBT-based converter requires 3.5 Ω for all the considered L_{out} and f_{sw}. Figure 3 shows the voltages and currents of one switching period (turn on and off) together with SOA region as indicated in the devices' datasheets. The Figure 3 resulted from the LTspice simulation considering the worst case with the highest one-leg current of 25 A, having for SiC-MOSFET-based inverter the L_{out} = 0.33 mH and f_{sw} = 100 kHz (Figure 3a), and for Si-IGBT-based converter L_{out} = 1.15 mH and f_{sw} = 30 kHz (Figure 3b).

Table 2. Main device parameters.

Parameters	Rohm SiC MOSFET SCT3080KLHR	SiC Schottky diode SCS220KG	Rohm Si IGBT RGS50TSX2DHR
V_{ds}	1200 V	1200 V	1200 V
I_{ds} (25 °C)	31 A	-	50 A
I_{ds} (100 °C)	22 A	20 A/133 °C	25 A
$R_{DS(on)}$ (25 °C)	80 mΩ	N/A	N/A
V_{CE-sat} (25 °C)	N/A	N/A	1.7 V
Q_g	60nC@18 V	N/A	67nC@15 V
V_{th}	2.7 V	N/A	6 V
V_{gs}	-4 to +22 V	N/A	±30 V
T_j	175 °C	175 °C	175 °C
P_{diss} (25 °C)	165 W	210 W	395 W
r_{jc}	0.7 °C/W	0.62 °C/W	0.38 °C/W

Table 3. Main simulation parameters.

SiC-MOSFET-Based Converter			Si-IGBT-Based Converter		
Inductance (mH)	Switching Frequency (kHz)	R_{g_ext} (Ω)	Inductance (mH)	Switching Frequency (kHz)	R_{g_ext} (Ω)
0.56	60	0	3.46	10	3.5
0.51	70	0	2.3	15	3.5
0.45	80	0	1.73	20	3.5
0.4	90	0	1.38	25	3.5
0.33	100	0	1.15	30	3.5

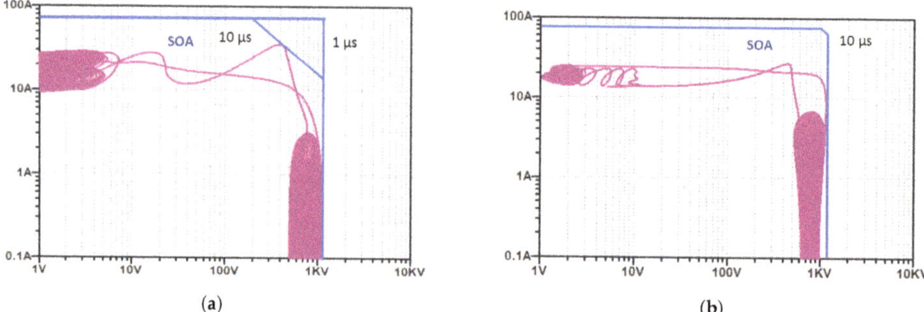

(**a**) (**b**)

Figure 3. Safe operating area: (**a**) SiC-MOSFET-based converter and (**b**) Si-IGBT-based converter.

As can be seen from the figure, the safe operating area is ensured for both cases. The same goes for the rest of the L_{out} and f_{sw} values.

4. Power Loss Analysis

The losses in the specific device are mainly the sums of two losses: the conduction and switching losses. The conduction losses of the IGBT can be calculated with the use of its dynamic on resistance $R_{on,IGBT}$ and zero on-state voltage V_{on}:

$$P_{con,IGBT} = V_{on}I_{av} + R_{on,IGBT}I_{rms}^{2} \tag{2}$$

where I_{av} and I_{rms} are the average and rms currents through the device.

On the other hand, the conduction losses of SiC-MOSFET can be evaluated by using its on-resistance $R_{DS(on)}$:

$$P_{con,MOSFET} = R_{DS(on)}I_{rms}^{2} \tag{3}$$

The conduction loss for the diode is based on its threshold voltage V_T and dynamic on resistance $R_{on,diode}$:

$$P_{con,diode} = V_T I_{av} + R_{on,diode}I_{rms}^{2} \tag{4}$$

The switching losses in the device depend on the switching frequency and the dissipated energies during turning on and turning off:

$$P_{sw,device} = f_{sw}\left(E_{on,device} + E_{off,device}\right) \tag{5}$$

where $E_{on,device}$ and $E_{off,device}$ are the device's dissipated energy during turning on and turning off.

4.1. Losses in the Interleaved dc–dc Converter

In the interleaved dc–dc converter, the conduction and switching losses can be analyzed by considering only one converter leg, since all the legs share the same losses. The output current has a DC value I_0 and a ripple component Δi_0. In the upper switch is circulating the current with the rms value [28]:

$$I_{rms,upper} = \sqrt{DI_0^2\left[1 + \frac{1}{12}\left(\frac{\Delta i_0}{I_0}\right)^2\right]} \tag{6}$$

where D is the duty cycle.

Similarly, the rms current of the bottom switch can be written as:

$$I_{rms,lower} = \sqrt{(1-D)I_0^2 \left[1 + \frac{1}{12}\left(\frac{\Delta i_0}{I_0}\right)^2\right]} \tag{7}$$

Basing on the Equations (6) and (7) for the conduction losses of SiC-MOSFET, Si-IGBT and the diodes we can write:

$$P_{cond,IGBT} = V_{on}D'I_0 + R_{on,IGBT}DI_0^2\left[1 + \frac{1}{12}\left(\frac{\Delta i_0}{I_0}\right)^2\right] \tag{8}$$

where $D' = (1-D)$.

$$P_{cond,MOSFET} = R_{DS(on)}DI_0^2\left[1 + \frac{1}{12}\left(\frac{\Delta i_0}{I_0}\right)^2\right] \tag{9}$$

$$P_{cond,diode} = V_TD'I_0 + R_{on,diode}D'I_0^2\left[1 + \frac{1}{12}\left(\frac{\Delta i_0}{I_0}\right)^2\right] \tag{10}$$

The switching losses can be evaluated as:

$$P_{sw,device} = f_{sw}E_{sw,device}\left(\frac{I_{avg}}{I_{ref}}\right)^{K_I}\left(\frac{V_{sup}}{V_{ref}}\right)^{K_V} \tag{11}$$

where I_{avg} is the average output current; V_{sup} is the device supply voltage (collector-emitter for IGBT, or drain-source for SiC-MOSFET); I_{ref} and V_{ref} are the respective current and voltage available in the datasheet, obtained from the switching loss measurement; K_I and K_V are the coefficients usually defined as in [29].

4.2. Inductor Losses

The losses in the inductor are the sums of winding losses and core losses. When it comes to winding losses, the DC resistance is associated with the dissipated power of the windings, but also phenomena such as skin effect and proximity effect. The latter two are associated with AC current components, and since the ripple of the current is minimized, can be neglected. The winding losses can be determined using [30]:

$$P_{L,w} = R_{dc}I_{rms}^2 \tag{12}$$

where I_{rms} is the rms current through the inductor.

The DC resistance can be calculated basing on the wire properties.

$$R_{DC} = \frac{\rho N mlt}{A_{winding}} \tag{13}$$

ρ being the specific copper resistivity, N the number of turns, mlt the mean length per turn available in the datasheet and $A_{winding}$ the winding's cross-sectional area.

The core losses can be approximated by the Steinmetz equation:

$$P_{L,core} = Kf^\alpha B_{pk}^\beta \tag{14}$$

where f is the frequency; B_{pk} is the peak flux density when applied for sinusoidal excitation; α and β are constants depending on the core material, magnetic induction and switching frequency operating range. Another way to estimate the core loss is by the use of core loss curves in the case of specific flux density (e.g., available in [31] for different core sizes and shapes).

The peak flux density can be calculated as [32]:

$$B_{pk} = \frac{E_{rms}10^8}{4.44 A_e N f_{sw}} \qquad (15)$$

where E_{rms} is the rms voltage across inductor (in V); A_e is the cross sectional area (in cm^2); and f_{sw} is the switching frequency (in Hz), while B_{pk} is expressed in gauss. The factor of 10^8 is due to the B_{pk} conversion from Tesla to gauss (1 Tesla = 10^4 gauss). Knowing B_{pk}, the core loss curves available in the core's datasheet can be utilized.

4.3. Double Pulse Test and PLECS Analysis

The simulation package PLECS offers the possibility to the user to merge the thermal and electrical design and provides the cooling solutions. The switching and conduction losses of the specific device are inserted by the user for each operating condition (forward current, blocking voltage, junction temperature) in terms of 3D look-up tables. In this way the simulation speed is not necessarily affected and the long thermal transient can be skipped by the steady-state analysis. The Cauer and Foster thermal networks can be utilized for the thermal description of the device.

The standard double pulse tests (DPT) have been conducted on the spice switch models presenting its realistic and dynamic behavior, as provided from the manufacturer's experimental tests. DPTs were done for each device listed in Table 2 in LTspice (Figure 4a with IGBT devices), in order to obtain the necessary switching and conduction losses of the device, used lateron in PLECS analysis. One example of the turn on losses obtained from LTspice DPT on a realistic switch model and implemented in PLECS is given in Figure 4b in the case of the 1200 V IGBT. In the same way, the losses for the SiC-MOSFET can also be determined and utilized.

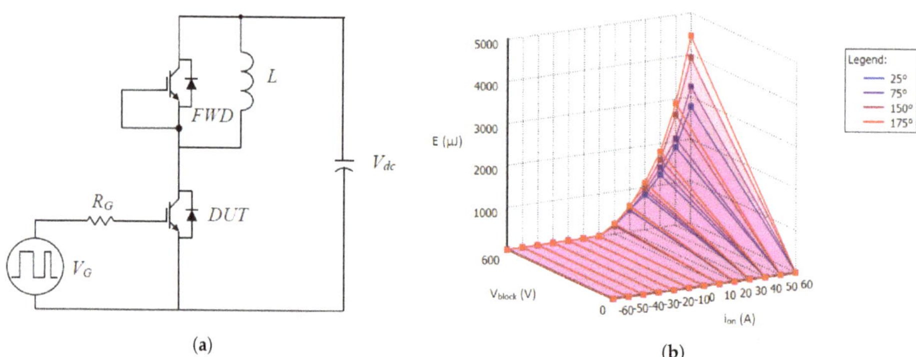

Figure 4. Double pulse test: (**a**) schematic of the double pulse test and (**b**) an example of IGBT's switching on losses for different temperatures.

After the loss descriptions had been added in PLECS and thermal network has been created, it was possible to select the appropriate heat sink. With the use of steady-state analysis tool, the respective converter losses together with the device junction, case temperatures and the heat sink temperature can be measured.

5. Results

Both LTspice and PLECS were used for the simulation analysis, as explained in Section 4.3. The device characterization in terms of conduction and switching losses has been done by double pulse tests on real device models in LTspice. In the next step, these losses connected to the different operating conditions were inserted in the look-up tables in PLECS and in this way it was possible to characterize each device for each switching period. The three-leg interleaved dc–dc converter (as shown in Figure 1) was implemented in PLECS. For the efficiency analysis the simpler converter model was considered, i.e., without the parasitics, as usually they can be neglected and are the same for the two converters. As resulted from the parasitics analysis, the external gate resistance of 0 Ω was used for double pulse test in the case of SiC-MOSFET device, considering only the internal resistance of the gate driver, i.e., 1.5 Ω. Instead, for the Si-IGBT device the external gate resistance of 3.5 Ω was used. The other simulation parameters are listed in Table 4. The L_{out}–f_{sw} pairs for each converter have been defined in Table 3 and were used here for simulations of different cases. The specific L_{out}–f_{sw} pairs were selected in order to keep the output current ripple within 5%. Cases with different output currents of 30, 45, 60 and 75 A were analyzed, corresponding to the 25%, 50%, 75% and 100% of the output power. The control reference was compared with three shifted carriers in order to obtain the pulse-width modulation (as explained in Section 2), providing a fixed switching frequency.

Table 4. Simulation parameters.

SiC-MOSFET Based Converter		Si-IGBT Based Converter	
Parameter (unit)	Value	Parameter (unit)	Value
V_{DC} (V)	800	V_{DC} (V)	800
R_{g_ext} (Ω)	0	R_{g_ext} (Ω)	3.5
R_{g_int} (Ω)	1.5	R_{g_int} (Ω)	1.5
D (%)	50	D (%)	50

5.1. Inductor Selection

The comparison of different core types given in Table 5 has been shown in Figure 5, for the two core types found as the best compromise for the total inductor losses and inductor volume. Two different types have been considered for both converters, iron (in turquoise) and alloy (in pink) powder core from [31]. In Figure 5 is given the comparison of total core losses, costs (obtained from the core manufacturer) and volume for the case of 25 A inductor current corresponding to the 75 A output current. In the case of SiC-MOSFET the difference between the volume of iron and alloy inductors is not very marked, except in the case of the highest switching frequency (0.33 mH–100 kHz) where the alloy core allows one to have an inductor with volume reduced to half (171 cm^3–85.5 cm^3).

For Si-IGBT-based converter there was more significant difference in the volume of iron and alloy core except for the 3.46 mH–10 kHz pair, where it was not possible to design the alloy powder core. When considering the lowest (10 kHz) and highest (30 kHz) switching frequencies, the alloy allowed 57% reduction in core volume.

Table 5. Comparison of different core types.

L–f_{sw} Pairs	Iron Core	Alloy Core
SiC-MOSFET-based converter		
0.56 mH, 60 kHz	**T400-14D** *	OP-521014-2
0.51 mH, 70 kHz	**T400-14D**	OP-521014-2
0.45 mH, 80 kHz	**T400-14D**	OP-521014-2
0.4 mH, 90 kHz	**T400-14D**	OP-521014-2
0.33 mH, 100 kHz	T400-14D	**SM-400026-2**
Si-IGBT-based converter		
3.46 mH, 10 kHz	**T650-14**	-
2.3 mH, 15 kHz	T650-14	**FS-650014-2**
1.73 mH, 20 kHz	T650-14	**OD-650026-2**
1.38 mH, 25 kHz	T650-14	**OD-601026-2**
1.15 mH, 30 kHz	T650-14	**FS-601014-2**

* In bold are given the inductors that later on are selected for further analysis.

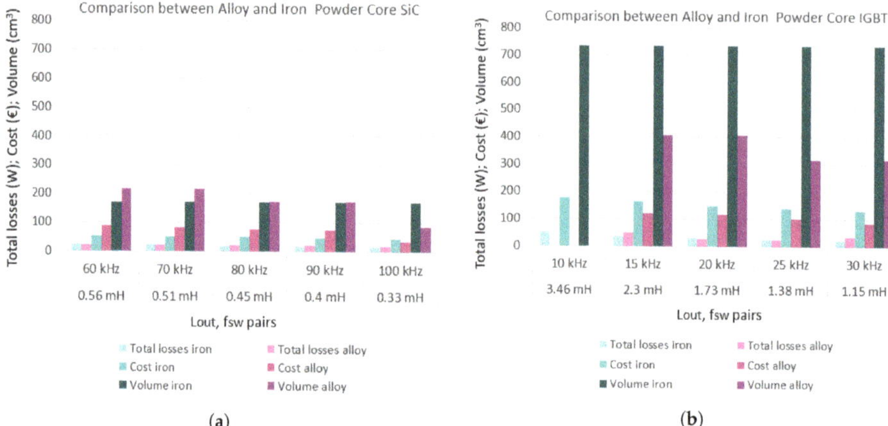

Figure 5. Comparison of cost, volume and total calculated losses for iron and alloy powder cores: (**a**) SiC-MOSFET-based converter; (**b**) Si-IGBT-based converter.

Evident also are the reductions in volume, losses and costs in both iron and alloy powder core when comparing Figure 5a,b for SiC-MOSFET-based converter, confirming the advantages of using the SiC-MOSFET devices. For the two highest switching frequencies (30 kHz for Si-IGBT and 100 kHz for SiC-MOSFET), the SiC-MOSFET-based converter can offer the reduction of 73% of the core volume for alloy core, and 77% for the iron core, having around 60% lower cost.

In this case, the best compromise in terms of total losses and volume has been taken into account. The selected inductors are given in bold in Table 5. In some cases the best choice is iron core, while in other cases it is alloy core. For the SiC-MOSFET-based converter the best solutions are almost always in iron, except for the case with the highest switching frequency (100 kHz), which also represents a significant working condition, allowing one to make the most of the advantages offered by SiC-MOSFET devices and significantly reduce weight and volume. On the other hand, for Si-IGBT the best solutions are always in alloy, except for the lowest switching frequency (10 kHz) with only possible solution in iron. However, this working condition is not very significant because it leads to higher volume and cost. Further on in the analysis, only the inductors given in bold were considered for the efficiency comparisons.

5.2. Power Loss and Efficiency Comparisons

In this section the power loss and efficiency comparison is presented. The three-leg interleaved dc–dc converter supplying a resistive load and implemented in PLECS has been used for this purpose. For each specific value of L_{out} the inductance with its winding resistance (as resulted from the inductor design shown in Table 6) has been used for the modeling. The converter losses have been determined directly from the PLECS simulation tool, while the inductor losses (for each specific inductor designed) have been calculated as described in Section 4.2.

In Figure 6 are given the dc–dc converter losses in the case of different output inductor–switching frequency pairs and output currents in the case of 50 °C heat sink temperature. In particular, four values of the output current have been considered: 30, 45, 60 and 75 A. The two converters show rather similar losses, even though the SiC-MOSFET converter's switching frequency is much higher than the one of Si-IGBT converter (60–100 kHz vs. 10–30 kHz). For the highest output current of 75 A, the SiC-MOSFET-based converter has 28%, 18% or 9% higher losses when compared to Si-IGBT-based converter for the switching frequencies below 80 kHz (for Si-IGBT below 20 kHz) respectively, but lower 1% for the 90kHz and 9% for 100 kHz, as shown in Figure 7a . Generally, all the losses show the same behavior, lower in case of Si-IGBT for the switching frequencies <20 kHz. More flat behavior of the losses for SiC-MOSFET-based converter can be noted, i.e., less variations with the change of L_{out}–f_{sw} pairs for the specific current.

In Figure 8 are given the total inductor losses for the selected inductors from Section 5.1. In particular, they resulted from the sum of inductor core and winding losses in the case of different inductors and output currents, as for converter losses. Different inductors have been designed, as shown in Table 6. The criteria for the design were the lowest total inductor losses, but also the lowest volume. The standard design practice has been applied; for example, the difference of the unloaded and loaded unductor was set to 10%, the cross section of the winding was carefully selected taking into account the minimum banding radius of the wire, the filling of the core window area was set to less than 50%, etc.

From the Table 6 can be noted 3.7 times lower inductor volume when comparing the cases with highest switching frequencies in the case of SiC-MOSFET-based converter. In Figure 8 the maximum losses for SiC-MOSFET-based converter of 24.9 W can be seen for the 0.56 mH–60 kHz pair and 75 A output current, while for the Si-IGBT-based converter the maximum is for 3.46 mH–10 kHz pair reaching 52.6 W at 75 A output current. The minimum losses for SiC-MOSFET-based converter are for 0.4 mH–90 kHz with 6.9 W in the case of lowest output current, while for Si-IGBT-based converter the minimum losses of 10.7 W are for 1.38 mH–25 kHz pair and the same value of the output current. In Figure 7b are given the inductor losses for the highest output current, i.e., 75 A. The losses of SiC-MOSFET-based converter are lower for all L_{out}–f_{sw} pairs, with the highest difference of 53% at the lowest switching frequency. Conveniently, the total one-leg losses are also depicted in Figure 7c, obtained as the sum of device losses and inductor losses per leg at 75 A output current. It is interesting to see that the Si-IGBT-based converter has higher losses in almost all cases, except the 1.73 mH–20 kHz case, where SiC-MOSFET-based converter shows 4% higher losses.

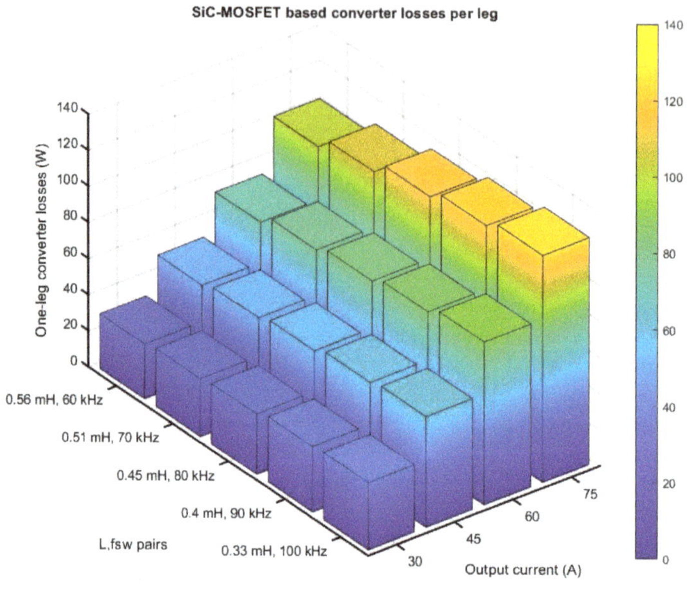

(**a**) SiC-MOSFET-based converter

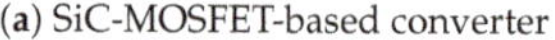

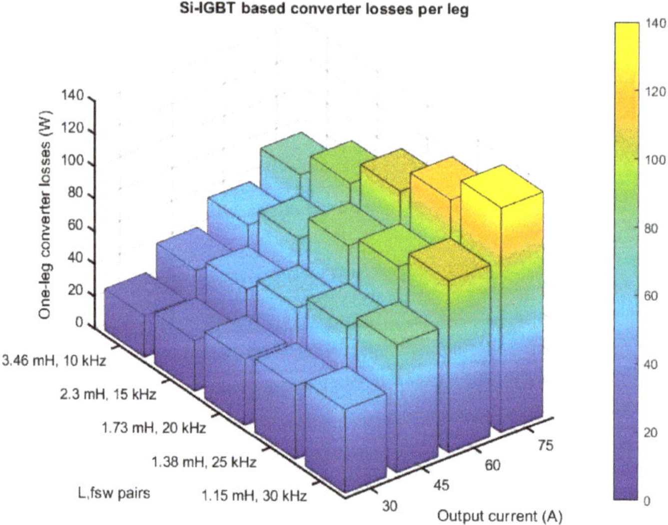

(**b**) Si-IGBT-based converter

Figure 6. Dc–dc converter losses per leg for different output inductors resulting from the simulation.

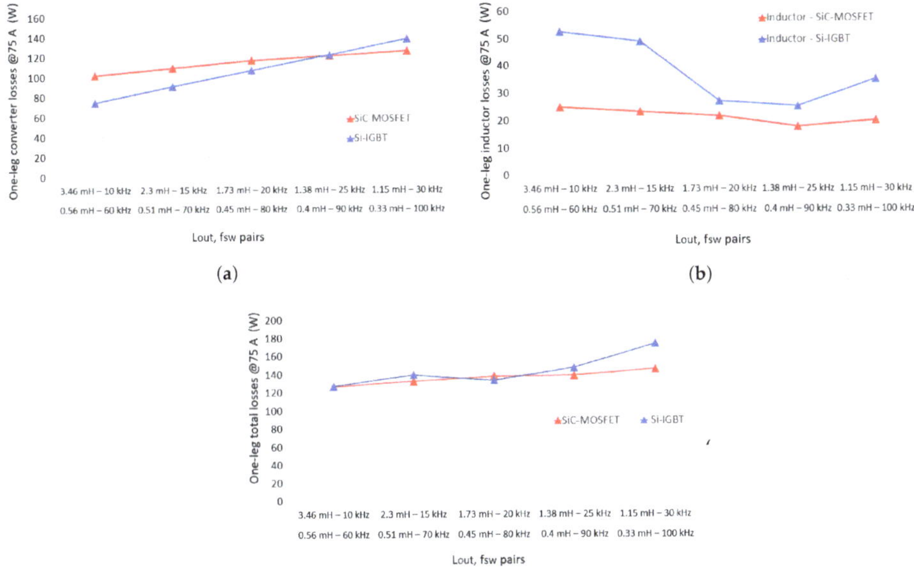

Figure 7. One-leg losses of SiC-MOSFET and Si-IGBT-based converters at 75 A output current: (**a**) simulated device losses, (**b**) analytical inductor losses, (**c**) total one-leg losses.

Table 6. Inductor design.

L_{out}, f_{sw} pairs	Core type	N	AWG	R_{dc} (Ω)	A_e (cm^2)	mlt	V_e (cm^3)
SiC-MOSFET Based Converter							
0.56 mH, 60 kHz	T400-14D	110	7	0.03082	6.85	14.62	171
0.51 mH, 70 kHz	T400-14D	105	7	0.02895	6.85	14.46	171
0.45 mH, 80 kHz	T400-14D	98	7	0.02649	6.85	14.24	171
0.4 mH, 90 kHz	T400-14D	92	6	0.02009	6.85	14.76	171
0.33 mH, 100 kHz	SM-400026-2	87	6	0.01522	3.5226	11.51	85.5
Si-IGBT Based Converter							
3.46 mH, 10 kHz	T650-14	212	6	0.07507	18.4	23.64	734
2.3 mH, 15 kHz	FS-650014-2	241	6	0.06565	9.87	18.09	407
1.73 mH, 20 kHz	OD-650026-2	155	5	0.03049	9.87	17.05	407
1.38 mH, 25 kHz	OD-601026-2	137	5	0.0273	8.8064	17.19	317
1.15 mH, 30 kHz	FS-601014-2	167	6	0.04266	8.8064	17.09	317

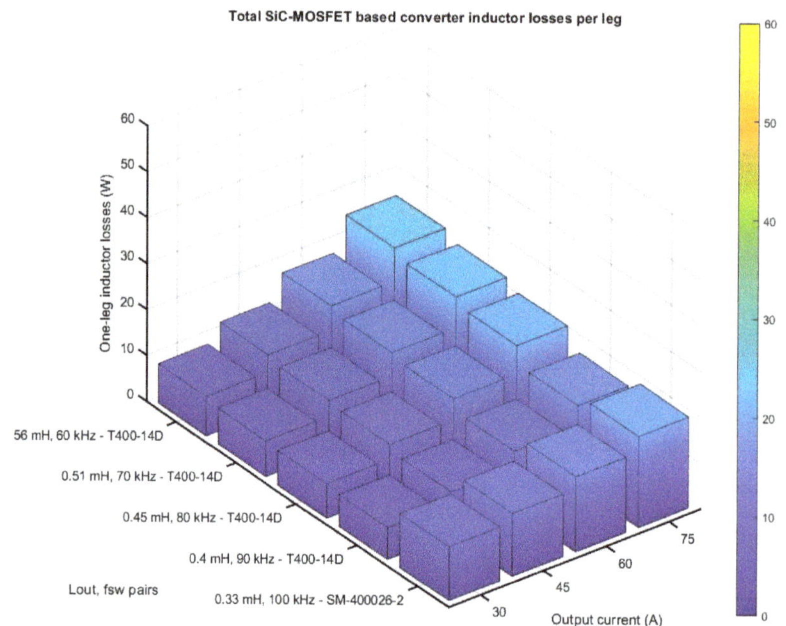

(**a**) SiC-MOSFET-based converter

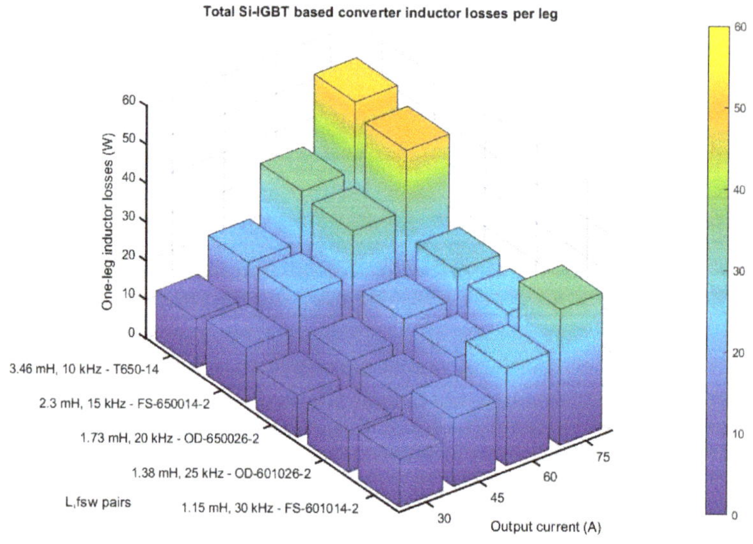

(**b**) Si-IGBT-based converter

Figure 8. Total inductor analytical losses per leg.

For the particular case in Figure 7c, where SiC-MOSFET-based converter shows slightly higher losses, the loss distribution has been analyzed for all the values of the output current. In Figure 9 is shown the distribution of the total losses (simulated device and analytical inductor) of one converter leg, and specifically for the 0.45 mH–80 kHz pair (SiC-MOSFET-based converter) and the 1.73 mH–20 kHz pair (Si-IGBT-based converter) in the case of 50 °C heat sink temperature. The figure also gives the comparison with the theoretical values, calculated as explained in Sections 4.1 and 4.2.

The total losses of SiC-MOSFET-based converter (dark pink) are higher only in the case of 75 A output current, as resulted also in Figure 7c, while for the other current values it shows lower losses compared to Si-IGBT converter (dark green). As for the losses of the devices in one converter leg, the switching losses which are also shown in the figure are slightly lower for the SiC-MOSFET-based converter, even though the switching frequency was four times higher. The total inductor loss (dotted traces) is slightly lower for SiC-MOSFET-based converter, having core losses invariant with the change of the output current, as expected. For the specific case of 75 A output current, the SiC-MOSFET-based converter shows elevated conduction losses, and this is the reason why it resulted in higher total losses. Moreover, also the good agreement between simulation results (first bar) and analytical results (second bar) for all cases can be observed.

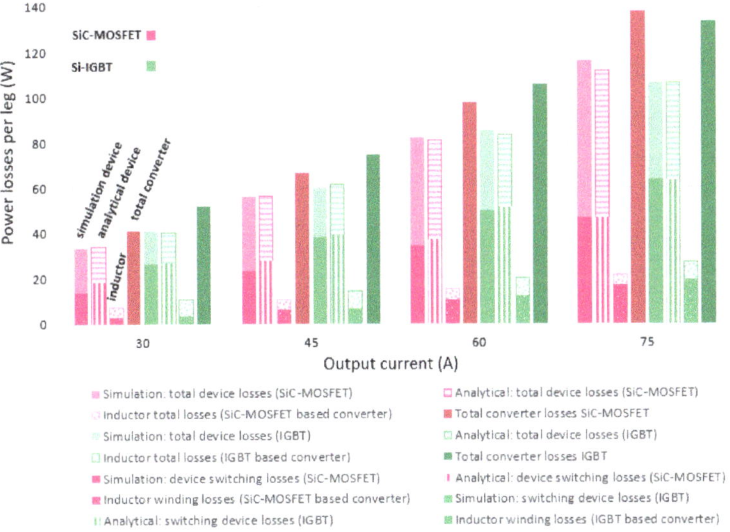

Figure 9. Loss distribution for 0.45 mH–80 kHz and 1.73 mH–20 kHz pairs and comparison with analytical values.

In Figure 10 are shown the total three-leg losses of the two converters, including the inductor losses resulting from realistic simulations (for power switches), and analytical for the different inductors. Generally, the SiC-MOSFET-based converter showed lower total losses (in pink), especially for the lower output currents. The highest losses can be noted for the Si-IGBT-based converter and 1.15 mH–30 kHz pair for all the values of the output current, due to high devices' losses. While the SiC-MOSFET-based converter has more uniform losses behavior with the change of switching frequency, this difference is more outlined in the Si-IGBT-based converter.

In Figure 11 are shown the efficiencies of the two three-leg converters for different L_{out}–f_{sw} pairs and output currents. For both converters, the efficiency is rather high, higher in the case of lower switching frequencies due to lower losses. The SiC-MOSFET-based converter shows higher efficiency (ranging between 98.3% and 98.9%), while in the case of Si-IGBT the efficiency ranges from 98% to 99%. The efficiency curves are closer to each other in the SiC-MOSFET when compared to Si-IGBT case, meaning that in the SiC-MOSFET case there is a lower variability of efficiency with the change of $L_{out}-f_{sw}$ pairs. In the worst case (30 A) there is a variability of about 0.3% in the case of SiC-MOSFET and 0.7% in the case of Si-IGBT. Therefore it can be concluded that by adopting the SiC-MOSFET solution, efficiency depends less on the choice of switching frequency and inductor.

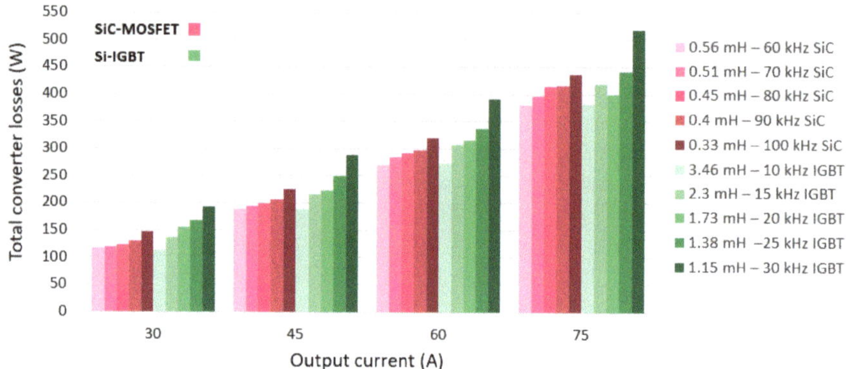

Figure 10. Total 3-leg converter loss for SiC-MOSFET-based and Si-IGBT-based converters.

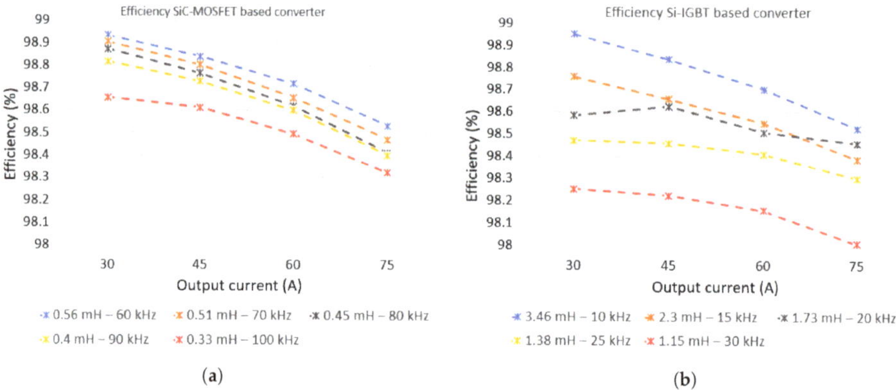

Figure 11. Efficiency comparison: (**a**) SiC-MOSFET-based converter and (**b**) Si-IGBT-based converter.

5.3. Heat Sink Volume

In order to conduct a throughout comparison of the two converters, the heat sink volume and cost should also be taken into account. The heat sink volume analysis was based on the power loss of the two converters at highest output current (75 A) and considering 50 °C heat sink temperature. In order to evaluate the heat sink volume, it is necessary to calculate the thermal resistance of the heat sink:

$$r_h = \frac{T_h - T_a}{P_{tloss}} \tag{16}$$

where T_h is the heat sink temperature, T_a is the ambient temperature and P_{tloss} is the total converter loss.

Once the r_h is calculated, it is possible to obtain the heat sink volume based on natural air convection [33]. The minimum heat sink volume can be obtained from the fitting function:

$$Vol_{heatsink} = 3263e^{-13.09r_h} + 1756e^{-1.698r_h} \tag{17}$$

where $Vol_{heatsink}$ is expressed in cm^3 and r_h in Ω. Equation (17) in [33] has the results of the curve fitting of various extruded naturally cooled heat sinks against heat sink thermal resistance.

Heat sink volume calculated from the curve fitting function for the two converters is presented in Figure 12, starting from the lowest switching frequency considered for both converters. A room temperature of 25 °C was selected as ambient temperature. The results show that the SiC-MOSFET-based converter has an increase of around 3% only in the case of 0.45 mH–80 kHz pair. In the other cases, the Si-IGBT-based converter has higher heat sink volume, with the highest difference of 7% in the case of highest switching frequency.

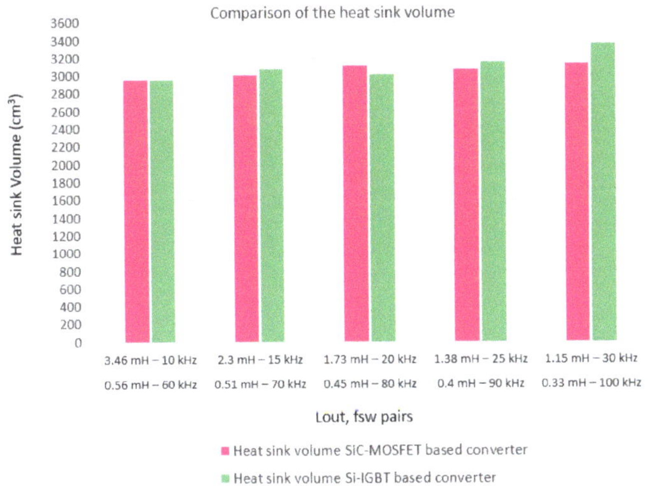

Figure 12. Comparison of the heat sink volume at 75 A output current and 50 °C heat sink temperature for the two converters.

5.4. Cost Comparison

This section presents the cost comparison of the two interleaved dc–dc converters, taking into account only the costs that are different for the two converters, i.e., the switching device costs, the heat sink costs and the inductor costs. The same cost was considered for the other components (for example gate driver, power cables, etc.), and therefore was not included in this comparison.

In Figure 13 is shown the comparison of the converter efficiency (3-legs, including inductors), the total volume (heat sink and inductor, in dm^3) and total converter cost (in € per kW). For the efficiency comparison, the power switch losses were taken directly from the realistic circuit model implemented in PLECS, while for the different inductors they were calculated analytically. Note that the costs in Figure 13 refer to the sample prices of the main parts' manufacturers in Europe. Having in mind that the price of SiC-MOSFET devices is 2–3 times the price of IGBT devices, and the costs of different inductors for the SiC-MOSFET-based converter generally being lower (due to lower volume), the goal of this analysis was to verify if the difference of the device cost can somehow be compensated with the lower inductor cost. Namely, Figure 13 confirms this fact, where the SiC-MOSFET-based converter shows lower total cost for all the L_{out}–f_{sw} pairs. The trend of the cost curve for the SiC-MOSFET-based converter is almost flat, with the lowest cost for the highest switching frequency, i.e., 100 kHz. The Si-IGBT-based converter shows more cost variation, with the lowest cost for the highest switching frequency as well. For the higher switching frequencies, the two costs are practically the same for each device.

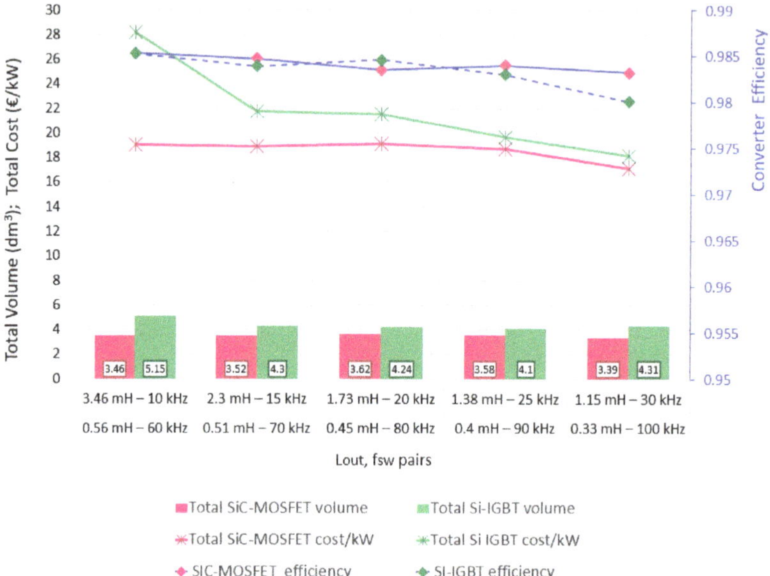

Figure 13. Efficiency, cost and total volume comparison of the two converters at 75 A output current and 50 °C heat sink temperature.

The total volume and efficiency confirm the advantage of using the SiC-MOSFET-based converter. The efficiency is almost always higher or equal to the one of the Si-IGBT-based converter. The only case in which the Si-IGBT-based converter showed higher efficiency was for 1.73 mH–20 kHz pair. The trend of the SiC-MOSFET efficiency curve is more flat, while the Si-IGBT-based converter shows higher variations with the lowest efficiency for 30 kHz switching frequency.

The real benefit can be seen in the total volume, where the lower volume can be noted for all the L_{out}–f_{sw} pairs in the SiC-MOSFET-based converter. For the highest switching frequency, SiC-MOSFET-based converter offers the lowest total volume. Generally, Figure 13 leads to the conclusion that the best working conditions for Si-IGBT-based converter could be medium switching frequencies (20 kHz and 25 kHz), where the cost, volume and efficiency differences are less marked, while the worst working conditions are the lowest switching frequencies, with elevated cost and volume difference. On the other hand, the real advantage of the SiC-MOSFET-based converter can be seen at high switching frequencies, where the cost is practically the same for the two converters (and lowest for all cases), efficiency remains higher and total volume is lower.

These considerations, together with the fact that the prices for SiC-MOSFET devices tend to reduce greatly over ther years, indicate the convenience of using SiC-MOSFET devices, especially given the fact that it offers lower total volume.

6. Conclusions

An investigation of the three-phase interleaved dc–dc topology applied to the EV fast chargers was presented in this paper. This configuration permits the use of classic and reliable three-phase power switch devices, featuring high reliability and modularity, making it particularly interesting for EV fast charging.

Two technologies were considered for the efficiency comparison: dc–dc interleaved converters with SiC-MOSFET power devices and Si-IGBT power devices, on a wide range of switching frequencies and output inductances. The comparison was made by imposing the same requirements on the ripple of the output current (5%), in order to reduce the battery stress and extend its life.

The two converters were considered with the optimal switching parameters, based on the analysis done on detailed realistic parasitic circuit model and experimental measurement. The goal was to have the safe operation and minimum losses for both converters. The output inductor design follows the same criteria. For the simulation analysis, the spice models describing the realistic and dynamic behavior of the power switches and based on the experimental tests were utilized and verified both with LTspice and PLECS simulation packages.

The results show the convenience of using SiC-MOSFETs for the three-phase interleaved dc–dc topology on the wide range of L_{out}–f_{sw} pairs. They also lead to the conclusion that the best working conditions for Si-IGBT-based converter could be medium switching frequencies (20 kHz and 25 kHz), where the cost, volume and efficiency differences are less marked, while the worst working condition is the lowest switching frequency, with elevated cost and volume difference. This is also the working condition where one could prefer the Si-IGBT solution due to its proven reliability and wide utilization by the industry. On the other hand, the real advantage of the SiC-MOSFET-based converter can be seen at high switching frequencies, where the cost is practically the same for the two converters, efficiency remains higher and total volume is lower.

Moreover, the analysis has shown that the higher price of SiC-MOSFET devices can be compensated for by the fact that the inductors utilized in the SiC-MOSFET-based converter have considerably lower costs due to the lower volume. In fact, the SiC-MOSFET-based converter showed lower costs for all L_{out}–f_{sw} pairs. However, this is likely to be changed with greater reductions in the price of SiC-MOSFET devices in upcoming years, making in this way the cost difference more significant.

Author Contributions: J.L.: conceptualization, data curation, formal analysis, investigation, methodology, software and writing—original draft. V.G.M.: formal analysis, investigation and writing—review and editing. G.L.C.: formal analysis and writing—review and editing. F.C.: formal analysis, writing—review and editing and funding acquisition. All authors have read and agreed to the published version of the manuscript.

Funding: This research was supported by the project "FURTHER—Future Rivoluzionarie Tecnologie per velivoli più Elettrici" Code ARS01_01283, funded by the Italian Ministry of University and Research within Programma Operativo Nazionale (PON) "Ricerca e Innovazione" 2014–2020.

Conflicts of Interest: The authors declare no conflict of interest.

References

1. Yilmaz, M.; Krein, P.T. Review of Battery Charger Topologies, Charging Power Levels, and Infrastructure for Plug-In Electric and Hybrid Vehicles. *IEEE Trans. Power Electron.* **2013**, *28*, 2151–2169. [CrossRef]
2. Ghosh, A. Possibilities and Challenges for the Inclusion of the Electric Vehicle (EV) to Reduce the Carbon Footprint in the Transport Sector: A Review. *Energies* **2020** *13*, 2602. [CrossRef]
3. Tran, V.T.; Sutanto, D.; Muttaqi, K.M. The state of the art of battery charging infrastructure for electrical vehicles: Topologies, power control strategies, and future trend. In Proceedings of the 2017 Australasian

Universities Power Engineering Conference (AUPEC), Melbourne, VIC, Australia, 19–22 November 2017; pp. 1–6.

4. Monteiro, V.; Ferreira, J.C.; Nogueiras Meléndez, A.A.; Couto, C.; Afonso, J.L. Experimental Validation of a Novel Architecture Based on a Dual-Stage Converter for Off-Board Fast Battery Chargers of Electric Vehicles. *IEEE Trans. Veh. Technol.* **2018**, *67*, 1000–1011. [CrossRef]

5. Musavi, F.; Eberle, W. Overview of wireless power transfer technologies for electric vehicle battery charging. *IET Elect. Power Appl.* **2014**, *7*, 60–66. [CrossRef]

6. Celli, G.; Soma, G.G.; Pilo, F.; Lacu, F.; Mocci, S.; Natale, N. Aggregated electric vehicles load profiles with fast charging stations. In Proceedings of the 2014 Power Systems Computation Conference, Wroclaw, Poland, 18–22 August 2014, pp. 1–7.

7. Barrero-González, F.; Milanés-Montero, M.I.; González-Romera, E.; Romero-Cadaval, E.; Roncero-Clemente, C. Control Strategy for Electric Vehicle Charging Station Power Converters with Active Functions. *Energies* **2019**, *12*, 3971. [CrossRef]

8. Dusmez, S.; Cook, A.; Khaligh, A. Comprehensive analysis of high quality power converters for level 3 off-board chargers. In Proceedings of the 2011 IEEE Vehicle Power and Propulsion Conference, Chicago, IL, USA, 6–9 September 2011; pp. 1–10. [CrossRef]

9. Channegowda, J.; Pathipati, V.K.; Williamson, S.S. Comprehensive review and comparison of DC fast charging converter topologies: Improving electric vehicle plug-to-wheels efficiency. In Proceedings of the 2015 IEEE 24th International Symposium on Industrial Electronics (ISIE), Buzios, Brazil, 3–5 June 2015; pp. 263–268.

10. Rivera, S.; Wu, B.; Kouro, S.; Yaramasu, V.; Wang, J. Electric Vehicle Charging Station Using a Neutral Point Clamped Converter with Bipolar DC Bus. *IEEE Trans. Ind. Electron.* **2015**, *62*, 1999–2009. [CrossRef]

11. Drobnic, K.; Grandi, G.; Hammami, M.; Mandrioli, R.; Ricco, M.; Viatkin, A.; Vujacic, M. An Output Ripple-Free Fast Charger for Electric Vehicles Based on Grid- Tied Modular Three-Phase Interleaved Converters. *IEEE Trans. Ind. Appl.* **2019**, *55*, 6102–6114. [CrossRef]

12. Hammami, M.; Viatkin, A.; Ricco, M.; Grandi, G. A DC/DC Fast Charger for Electric Vehicles with Minimum Input/Output Ripple Based on Multiphase Interleaved Converters. In Proceedings of the IEEE ICCEP International Conference on Clean Electrical Power, 2–4 July 2019; pp.187–192.

13. Hwu, K.I.; Yau, Y.T. An Interleaved AC–DC Converter Based on Current Tracking. *IEEE Trans. Ind. Electron.* **2009**, *56*, 1456–1463. [CrossRef]

14. Zhang, D.; Wang, F.; Burgos, R.; Lai, R.; Boroyevich, D. DC-link ripple current reduction for paralleled three-phase voltage-source converters with interleaving. *IEEE Trans. Power Electron.* **2011**, *26*, 1741–1753. [CrossRef]

15. Huang, Y.; Tan, S.; Hui, S.Y. Multiphase-Interleaved High Step-Up DC/DC Resonant Converter for Wide Load Range. *IEEE Trans. Power Electron.* **2019**, *34*, 7703–7718. [CrossRef]

16. Devaraj, E.; Joseph, P.K.; Karuppa Raj Rajagopal, T.; Sundaram, S. Renewable Energy Powered Plugged-In Hybrid Vehicle Charging System for Sustainable Transportation. *Energies* **2020**, *13*, 1944. [CrossRef]

17. Tan, L.; Zhu, N.; Wu, B. An Integrated Inductor for Eliminating Circulating Current of Parallel Three-Level DC–DC Converter-Based EV Fast Charger. *IEEE Trans. Ind. Electron.* **2016**, *63*, 1362–1371. [CrossRef]

18. Biela, J.; Schweizer, M.; Waffler, S.; Kolar, J.W. Evaluation of potentials for performance improvement of inverter and DC–DC converter systems by SiC power semiconductors. *IEEE Trans. Ind. Electron.* **2011**, *58*, 2872–2882. [CrossRef]

19. Anthon, A.; Zhang, Z.; Andersen, M.A.E.; Holmes, D.G.; McGrath, B.; Teixeira, C.A. The Benefits of SiC mosfets in a T-Type Inverter for Grid-Tie Applications. *IEEE Trans. Power Electron.* **2017**, *32*, 2808–2821. [CrossRef]

20. Zhang, D.; He, J.; Madhusoodhanan, S. Three-Level Two-Stage Decoupled Active NPC Converter with Si IGBT and SiC MOSFET. *IEEE Trans. Ind. Appl.* **2018**, *54*, 6169–6178. [CrossRef]

21. Loncarski, J.; Monopoli, V.G.; Leuzzi, R.; Ristic, L.; Cupertino, F. Analytical and Simulation Fair Comparison of Three Level Si IGBT Based NPC Topologies and Two Level SiC MOSFET Based Topology for High Speed Drives. *Energies* **2019**, *12*, 4571. [CrossRef]

22. Loncarski, J.; Ricco, M.; Monteiro, V.; Monopoli, V.G. Efficiency Comparison of a dc-dc Interleaved Converter Based on SiC-MOSFET and Si-IGBT Devices for EV Chargers. In Proceedings of the 14th International Conference on Compatibility, Power Electronics and Power Engineering (CPE-POWERENG), Setubal, Portugal, 8–10 July 2020; pp. 517–522.
23. Stefanskyi, A.; Starzak, Ł.; Napieralski, A. Review of commercial SiC MOSFET models: Validity and accuracy. In Proceedings of the 24th International Conference Mixed Design of Integrated Circuits and Systems, Bydgoszcz, Poland, 22–24 June 2017; pp. 488-493.
24. Rosa, E.B. The Self and Mutual Inductances of Linear Conductors. In *Bulletin of the Bureau of Standards*; Dept. of Commerce and Labor, Bureau of Standards, U.S. Govt. Print. Off.: Washington, DC, USA, 1908; pp. 301–344.
25. SCT3080KLHR Product Information. Available online: https://www.rohm.com/products/sic-power-devices/sic-mosfet/sct3080klhr-product (accessed on 9 July 2020).
26. SCS220KG Product Information. Available online: https://www.rohm.com/products/sic-power-devices/sic-schottky-barrier-diodes/scs220kg-product (accessed on 9 July 2020).
27. RGS50TSX2DHR Product Information. Available online: https://www.rohm.com/products/igbt/field-stop-trench-igbt/rgs50tsx2dhr-product (accessed on 9 July 2020).
28. Femia, N.; Petrone, G.; Spagnuolo, G.; Vitelli, M. *Power Electronics and Control Techniques for Maximum Energy Harvesting in Photovoltaic Systems*; CRC Press: Boca Raton, FL, USA, 2013.
29. Semikron: Application Manual Power Semiconductors. Available online: https://www.semikron.com/service-support/downloads/ (accessed on 5 June 2020).
30. Erickson, R.W.; Maksimovic, D. *Fundamentals of Power Electronics*, 2nd ed.; Springer: Berlin/Heidelberg, Germany, 2001.
31. MicroMetals. Available online: https://www.micrometals.com (accessed on 5 June 2020).
32. Power Conversion and Line Filter Applications (Micro Metals Iron Powder Core Application Note, Issue L, Feb. 2007.). Available online: https://www.tme.eu/Document/82fc93ef93d05a020041f9203a13131a/MICROMETALS-Tseries.pdf (accessed on 9 July 2020).
33. Gurpinar, E.; Castellazzi, A. Single-Phase T-Type Inverter Performance Benchmark Using Si IGBTs, SiC MOSFETs, and GaN HEMTs. *IEEE Trans. Power Electron.* **2016**, *31*, 7148–7160. [CrossRef]

Review

State-Space Modeling Techniques of Emerging Grid-Connected Converters

Fabio Mandrile [1], Salvatore Musumeci [1,*], Enrico Carpaneto [1], Radu Bojoi [1], Tomislav Dragičević [2] and Frede Blaabjerg [3]

[1]   Energy Department "Galileo Ferraris" (DENERG), Politecnico di Torino, Corso Duca degli Abruzzi 24, 10129 Torino, Italy; fabio.mandrile@polito.it (F.M.); enrico.carpaneto@polito.it (E.C.); radu.bojoi@polito.it (R.B.)

[2]   Department of Electrical Engineering, Technical University of Denmark, 2800 Kgs. Lyngby, Denmark; tomdr@elektro.dtu.dk

[3]   Department of Energy Technology, Aalborg University, 9220 Aalborg, Denmark; fbl@et.aau.dk

*   Correspondence: salvatore.musumeci@polito.it

Received: 17 August 2020; Accepted: 11 September 2020; Published: 15 September 2020

Abstract: In modern power electronics-based power systems, accurate modeling is necessary in order to analyze stability and the interaction between the different elements, which are connected to it. State space modeling seems a valid approach to study the modes of a certain system and their correlation with its states. Unfortunately, this approach may require complicated calculations and it is difficult to model advanced or emerging control techniques for grid-tied converters, such as cascaded controllers (e.g., voltage and current) and virtual synchronous generators (VSGs). Moreover, this approach does not allow an easy reconfiguration of the modeled system by adding, removing of modifying certain elements. To solve such problems, this paper presents a step-by-step approach to the converter modeling based on the Component Connection Method (CCM). The CCM is explained in detail and a practical example is given, by modeling one exemplary VSG model available in the literature. The obtained model is finally validated experimentally to demonstrate the practical accuracy of such approach.

Keywords: Component Connection Method; power electronics-based systems; stability analysis; state-space methods; virtual synchronous generators

---

1. Introduction

The transition to power systems involving more power electronics-based converters represents a challenge today. An increase in the penetration of inverter-interfaced renewable energy sources (such as wind and sun) may lead to instabilities within the power system [1,2]. For this reason, accurate converter models, taking into consideration both the converter hardware parameters and its control algorithm, will be necessary. Moreover, some grid codes are now requiring dynamic models on different degree of details for simulation and analysis at the power system level [3,4].

Several techniques are presently available for this purpose. The most popular are frequency domain analysis using impedance models and the eigenvalue analysis using state-space models [5]. Regarding the impedance models, various approaches are available in the literature [6–11], as well as experimental procedures [12] to obtain the black-box frequency response of a converter without knowing its internal parameters. Regarding the system-level analysis, the Generalized Nyquist Criterion can be used [13] if the individual equivalent impedance models are available. However, impedance models fail to provide an immediate comprehension of the poles in the system and how they are influenced by the system parameters. Moreover, by observing the Bode plot of the equivalent impedance of a converter, the effect of some poles can be canceled by corresponding zeros of the system.

On the contrary, the state-space approach has the advantage of clearly identifying the modes of the system under study and their correlation both to its states (e.g., controller, filter) using participation factors [14] and to the system parameters using sensitivity analysis. Moreover, the state-space model can provide an immediate feedback of the different poles time constants, without incurring the effects of zero-pole cancellation that can be present in the impedance model representation.

Traditionally, compared to synchronous generators, grid-connected inverters have been working with much smaller time constants. These time constants are related to both the physical hardware of the converters (i.e., grid side filters) and the digital controllers. However, to obtain a higher integration of renewable energy sources, inverters may be required to provide ancillary services to the grid, in order to maintain its stability and quality (frequency, voltage and harmonic content). These services typically operate with much slower time constants, in the order of magnitude of the line frequency or less. For these reasons, a multi time scale analysis of the system may be necessary. Also, simplified models may be desirable, focusing only on certain phenomena, to reduce the computational burden of otherwise extensive simulations to assess the stability and the behavior of the system.

However, state-space modeling is a very difficult task when dealing with complex systems internally interconnected. An increase in the number of state variables, inputs and outputs of the system leads to complicated matrix expressions. Moreover, another disadvantage is its low flexibility: if components are added, removed or modified, the modeling process must be repeated from the beginning. This disadvantage is particularly evident when modeling converters equipped with advanced cascaded or more emerging controllers. Such controllers may be implemented to provide additional features beyond the traditional current or voltage control of grid-connected converters and are an enabling technology to integrate more renewable energy sources in the future power systems.

An example of outer advanced controller are the virtual synchronous generators (VSGs), which can provide virtual inertia, harmonic compensation and reactive support during voltage dips [15–22].

The literature proposes some examples of state-space modeling of such controllers [18], but following a monolithic approach, i.e., deriving the complete model at once. This kind of approach is feasible, but it lacks of flexibility, modularity and scalability (i.e., if something changes, the model must be derived again). Moreover, the number of state variables involved dramatically increases the order of the obtained system. This complexity leads to a larger modeling effort and it is prone to errors. Therefore, troubleshooting cannot be easily done during the derivation of the model.

A valid methodology to tackle such issues is the Component Connection Method (CCM). This method solves the lack of modularity and simplifies the modeling procedure, reducing both the analytical effort and the probability of making mistakes during the derivation, by separately modeling the single components of the system and then linking them using sparse interconnection matrices, as shown in Figure 1.

To provide a global and practical perspective on this technique, this paper reviews the requirements of power electronics modeling in power systems and the most recent applications of CCM to this matter. Moreover, it contributes with a generalized step-by-step Component Connection Method (CCM)-based modeling approach of the individual converter unit, considering both the hardware and the control part. Compared to the classical CCM theory and existing applications, purely algebraic blocks are introduced in this paper and it is described how to include them in the CCM in a completely general way. Such blocks have no physical meaning in power systems applications, but they simplify the modeling of power electronics controllers (e.g., calculating the current references starting from the power setpoints and the grid voltage).

This paper is organized as follows. In Section 2 the state of the art of CCM applied to power electronics-based power systems is reviewed, along with its most recent applications. Section 3 presents a step-by step procedure of CCM-based modeling. A practical particular example of such method by modeling the Simplified Virtual Synchronous Compensator (S-VSC) [23], is given in Section 4. Section 5 presents some analyses, which can be performed thanks to the state-space model and the experimental validation of the presented case study. Finally, the conclusions are provided in Section 6.

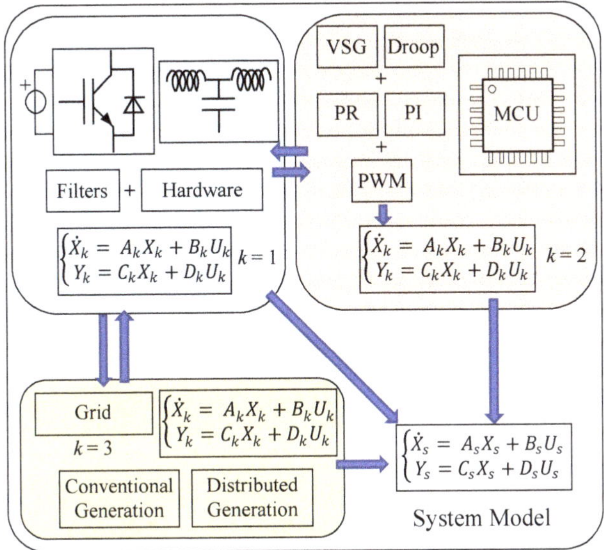

Figure 1. Modular state-space modeling of a grid connected converter including controllers.

2. State-Space Modeling Literature

With the spread of distributed generation and microgrids based on power electronics converters, accurate and efficient modeling techniques have become necessary. Such methods must guarantee flexibility (i.e., being easily applicable to different systems), scalability (i.e., can model more than one converter and can merge multiple grid sub-portions) and various detail level, depending on the specific analysis needed.

For this purpose, various levels of complexity and detail have been proposed during the years. At first, the focus was on low frequency subsynchronous oscillations, neglecting the faster behavior of current/voltage controllers [24]. In such cases, the inverter was considered as an ideal voltage source, controlled in amplitude and frequency by active and reactive droop controllers. However, such level of simplification is not always accurate in predicting instabilities or poorly damped resonances occurring at higher frequencies (innermost current or voltage control bandwidth and grid-side filters resonance frequency). Therefore, more accurate models have been proposed [25,26].

Given these requirements and summarizing the benefits of state-space modeling briefly mentioned in Section 1:

- The modes of the system under study are evident in their frequency and damping. Therefore, poorly damped oscillations can be easily identified;
- The interaction among the modes and the states can be analyzed by means of participation factors. The states associated with the most critical modes are then identified;
- A reduced order model of the system can be obtained, by neglecting the dynamic behavior of some of its parts. This allows faster simulations without a decrease in the quality of the results.

CCM is a valid modeling approach. In fact, the main advantage of CCM is the possibility of decomposing an articulated system into its fundamental blocks, which can then be individually modeled, and the possibility of reconfiguring the system by simply adding new blocks or just modifying the interconnection matrices. Such advantage is especially beneficial when modeling a single converter unit. This allows an easy an quick derivation of the state-space model of the converter, even when a single or more control layers (e.g., current controller) or hardware component (e.g., grid side filter) are modified in a later stage of the design.

CCM has been traditionally used at power system level to analyze multi-node networks with several conventional generating units or loads starting from [27]. Later, it has been specifically used to model the early grid applications of power electronics: static var compensators and HVDC transmission systems [28,29]. This method has gained renewed interest in the last decade, when it was applied to modern power electronics-based power systems. First, in 2014 CCM was applied to the specific modeling of wind farms and their connection to the main grid [8,30,31]. The same research evolved then in CCM modeling applied to multi converter power system, being able to include highly detailed inverter models (i.e., including accurate control loops and digital delays models) in wider inverter-based power systems.

CCM is currently used to analyze the stability and the interaction among conventional (i.e., phase locked loop (PLL)-based) and new controllers (i.e., Synchronverters) for renewable generation [32–34]. In these works, the CCM is the first stage of a more advanced state-space μ-analysis [34] to consider the uncertainties of the modeled plants.

CCM is still also currently adopted in power system studies, to include inverter-based grid nodes into larger scale studies (e.g., Subsynchronous torsional interactions [35]). In such cases, its flexibility is exploited in order to simplify the modeling procedure neglecting aspects at frequency out of the range of interest (e.g., digital control delay). The history of CCM is schematically summarized in Figure 2.

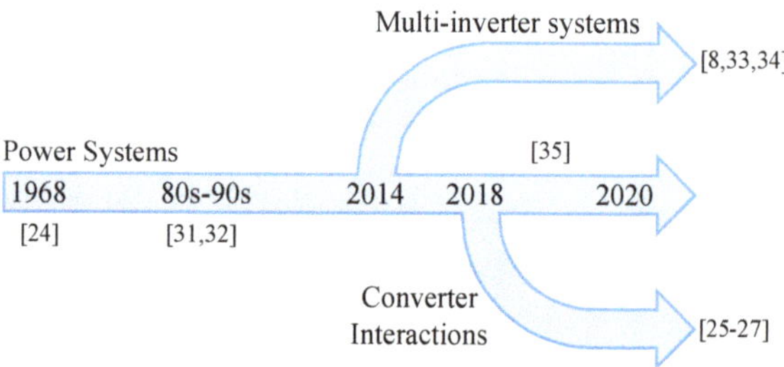

Figure 2. Applications and trends of CCM in time.

3. Generalized CCM Modeling Approach

This section considers a step-by-step procedure to model a grid-connected three-phase two level inverter, interfaced to the grid by means of an LCL filter and is depicted in Figure 3. The DC supply of the inverter is considered to be an ideal voltage source. Thanks to the modular approach guaranteed by the CCM, a more accurate model of the DC source, or even the model of a more complex DC side, e.g., a DC microgrid [36,37], can be easily included. The grid is modeled as a Thévenin equivalent circuit.

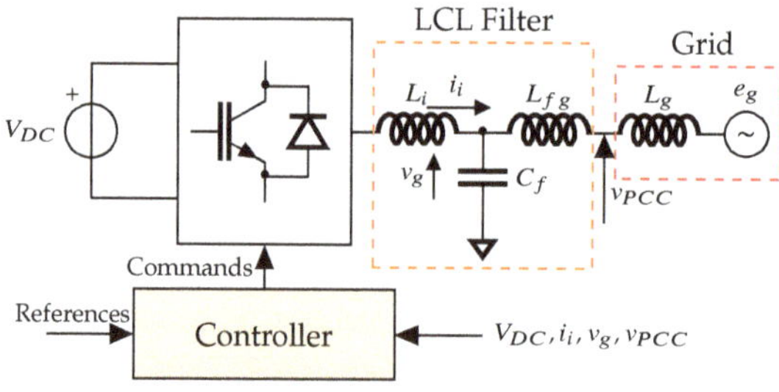

Figure 3. Diagram of the system to be modeled.

The CCM consists of the modular analysis of the system under modeling. First, the single components are identified and linearized around the chosen operating point. Then, the connections between the components are defined by means of sparse interlinking matrices. Finally, the state-space model of the overall system is obtained by linear algebra calculation.

This process is performed by the following general step-by-step procedure described below:

1. Define the system inputs U_s and outputs Y_s. Examples of inputs are external disturbances such as references variations (e.g., active and reactive power) and grid faults (i.e., frequency, voltage amplitude or phase variations). The outputs of the system can be chosen arbitrarily, depending on the aspect under study. Examples are the power (or current) injected by the inverter and the measured grid frequency using PLLs or VSGs.

2. Identify the single components. Two kinds of components can be used: dynamic (with state variables) and algebraic (without state variables). In general, n dynamic components and m algebraic components are identified and linearized.

3. Each k-th linearized dynamic component can be written in the following state-space form:

$$\begin{cases} \dot{X}_k &= A_k X_k + B_k U_k \\ Y_k &= C_k X_k + D_k U_k \end{cases} \qquad k = 1 \ldots n \qquad (1)$$

where X_k represents the state variables vector of the block, U_k the inputs of the component and Y_k its outputs. The matrices A_k, B_k, C_k and D_k are the component state-space matrices. Examples of the dynamic components are LCL filters and integral regulators.

4. Algebraic components are defined to simplify the modeling procedure. In a traditional state-space representation, only dynamic components are defined. However, this distinction can be very useful when it comes to the modeling of a controller. Many algebraic calculations, such as divisions, lead to complicated linearized expressions and it can be convenient to simplify the process by defining such algebraic components. They do not have state variables, but only inputs U_k and outputs Y_k related as:

$$Y_k = D_k U_k \qquad k = n+1 \ldots n+m \qquad (2)$$

Droop controllers, sums and divisions to calculate the current reference from the active and reactive power references are examples of algebraic blocks.

5. The vector of the state variables of the global system X_s is the aggregate of the X_k state variable vectors of the single components. The list of the components inputs U and the list of the components outputs Y can be defined in an equivalent way as follows:

$$\begin{aligned} X_s &= [X_1 \ldots X_n]' \\ U &= [U_1 \ldots U_{n+m}]' \\ Y &= [Y_1 \ldots Y_{n+m}]' \end{aligned} \qquad (3)$$

6. An aggregated model is defined by composing the individual system matrices as follows:

$$\begin{cases} \dot{X}_s &= A_a X_s + B_a U \\ Y &= C_a X_s + D_a U \end{cases} \qquad (4)$$

where:

$$A_a = \begin{bmatrix} A_1 & 0 & \cdots & 0 \\ 0 & A_2 & \cdots & 0 \\ \vdots & \vdots & \ddots & \vdots \\ 0 & 0 & \cdots & A_n \end{bmatrix}$$

$$B_a = \begin{bmatrix} B_1 & 0 & \cdots & 0 & \overbrace{0 \quad \cdots \quad 0}^{\text{number of algebraic blocks inputs}} \\ 0 & B_2 & \cdots & 0 & 0 & \cdots & 0 \\ \vdots & \vdots & \ddots & \vdots & \vdots & \ddots & \vdots \\ 0 & 0 & \cdots & B_n & 0 & \cdots & 0 \end{bmatrix}$$

$$C_a = \left.\begin{bmatrix} C_1 & 0 & \cdots & 0 \\ 0 & C_2 & \cdots & 0 \\ \vdots & \vdots & \ddots & \vdots \\ 0 & 0 & \cdots & C_n \\ 0 & \cdots & \cdots & 0 \\ \vdots & \vdots & \ddots & \vdots \\ 0 & \cdots & \cdots & 0 \end{bmatrix}\right\} \begin{array}{l} \text{number of algebraic} \\ \text{blocks outputs} \end{array}$$

$$D_a = \begin{bmatrix} D_1 & 0 & \cdots & 0 \\ 0 & D_2 & \cdots & 0 \\ \vdots & \vdots & \ddots & \vdots \\ 0 & 0 & \cdots & D_{n+m} \end{bmatrix}.$$

$$\tag{5}$$

In (5), it must be noted that the B_a and C_a matrices must be extended by adding null elements to compensate for the algebraic blocks. In particular, a number of null columns equal to the total number of inputs of all the algebraic blocks must be attached to the B_a matrix. Also several null rows equal to the sum of the algebraic block outputs must be attached to the C_a matrix.

7. Define the connection matrices, to connect the single components.

These matrices T are sparse and connect the inputs and outputs of the system as follows:

$$\begin{aligned} U &= T_{uy}Y + T_{us}U_s \\ Y_s &= T_{sy}Y + T_{ss}U_s \end{aligned} \tag{6}$$

8. The global linearized system state-space model can be obtained as:

$$\begin{cases} \dot{X}_s &= A_s X_s + B_s U_s \\ Y_s &= C_s X_s + D_s U_s \end{cases} \tag{7}$$

where:

$$\begin{aligned} A_s &= A_a + B_a T_{uy} W C_a \\ B_s &= B_a T_{uy} W D_a T_{us} B_a T_{us} \\ C_s &= T_{sy} W C_a \\ D_s &= T_{sy} W D_a T_{us} + T_{ss} \\ W &= \left(I - D_a T_{uy}\right)^{-1}. \end{aligned} \tag{8}$$

when the system structure is modified, U and Y must be modified according to the new blocks, as well as the connection matrices.

As mentioned in the introduction, the CCM is useful when performing a multi time scale analysis of a system. The reason behind this kind of analysis is that the most recent controllers, enabling power electronic converters provide ancillary services to the grid (e.g., VSGs) typically work on different time scales. The inner control loops (voltage, current) are operating at the time scale of the switching frequency. On the other hand, the outer level controllers, managing for example the virtual inertia, operate with time constants in the order of magnitude of seconds. To perform a simplified analysis with a traditional state-space modeling approach, the system has to be completely remodeled, meaning time loss and possibility of introducing errors. The CCM solves this problem: the user can easily exclude some levels of the system by substituting them with purely algebraic blocks. This way the computational burden for simulations is reduced, while preserving the correctness of the analysis at the time scale of interest. This simplification affects only the state variables of the k-th excluded block, but does not affect its inputs and outputs. Therefore, a dynamic block can be transformed into an algebraic block and the global system is easily derived again, without any need to change the connection matrices. The poles relative to the neglected blocks will not be present in the global system anymore. The pole map of the system and a participation factors analysis, which will be described in Section 5, are a good starting point to decide which blocks can be neglected to perform which study, since they give a clear and straightforward view of the time constant of each pole and which blocks are related to them.

Moreover, the CCM is flexible and solves also the issues related to the comparison or improvements of the inverter control and hardware. In fact, CCM decompose an articulated system into its fundamental blocks, which can then be individually modeled, and allows the reconfiguration of the system by simply adding new blocks or just modifying the interconnection matrices. Such advantage is especially beneficial when modeling a single converter unit. The benefits are especially evident in comparison to the traditional monolithic modeling approach, which has to be repeated again when any part of the system is modified. This flexibility of CCM allows an easy and quick derivation of the new state-space model of the converter, even when a single or more control layers (e.g., current controller) or hardware component (e.g., grid side filter) are modified in a later stage of the design.

A practical example of the CCM superiority can be done by comparing it with the traditional state-space modeling. The literature [18] gives examples of VSG modeling with traditional techniques. In this example the complete state-space model of a grid-connected converter controlled with a cascaded VSG is described. However, the equivalent model is obtained directly, with a monolithic approach. The result is a 19×19 state variable matrix, with no easy identification of which part of this matrix corresponds to which block of the converter. With the CCM, as demonstrated by the approach of (5), this is not the case. Each functional block of the system under study corresponds to a well defined matrix, which can be modified in a later stage. For example, in case a different current control strategy is implemented in a later stage of the design, its inputs and outputs will not probably change, but only its internal structure. Several controllers can be therefore compared, as well as different approximations of the delay of the digital controller.

4. Modeling of the S-VSC

In this section, an example of the application of the CCM is given. The system under study is the Simplified Virtual Synchronous Compensator (S-VSC). This model has been chosen as a representative of a VSG-based controller providing ancillary services to the grid. However, CCM can be applied to model any other controller available in the literature, by being a general modeling approach.

The system consists of the inverter hardware part, its control (current controller and S-VSC) and the connection to the equivalent grid, as shown in Figure 4. The system is modeled in per unit with V_b base voltage, S_b base power and ω_b base angular speed.

More details on this model are available in the literature [23].

4.1. Defining System Inputs and Outputs

First, the inputs and the outputs of the system are defined. The system inputs U_s are defined to model the possible electrical perturbations that can influence the system. In particular, step variations in the active and reactive power references (ΔP_{ext}^*, ΔQ_{ext}^*), grid frequency $\Delta \omega_g$, grid voltage amplitude ΔE_g and grid voltage phase angle $\Delta \phi_g$ variations. This choice allows full testing of the system under not-rated operating conditions.

The system outputs Y_s are arbitrarily selected, depending on the quantities of interest. In this example, they have been chosen as follows:

$$
\begin{aligned}
U_s &= \left[\Delta P_{ext}^*, \Delta Q_{ext}^*, \Delta \omega_g, \Delta E_g, \Delta \phi_g\right]'_{5\times 1} \\
Y_s &= \left[\Delta P_i, \Delta Q_i, \Delta \omega_r, \Delta \delta, \Delta v_g^d, \Delta v_g^q\right]'_{6\times 1}
\end{aligned}
\tag{9}
$$

where ΔP_i and ΔQ_i are the active and reactive powers injected by the inverter.

4.2. Component Identification

The components are identified as shown in Figure 4. $n = 4$ dynamic blocks and $m = 2$ algebraic blocks are defined as follows:

- Dynamic components: LCL filter, Inverter control loops (PI and delay model), S-VSC electrical part (stator and damper), S-VSC power loops (mechanical part and excitation control);
- Algebraic components: Power reference calculation (power to current) and grid perturbation model.

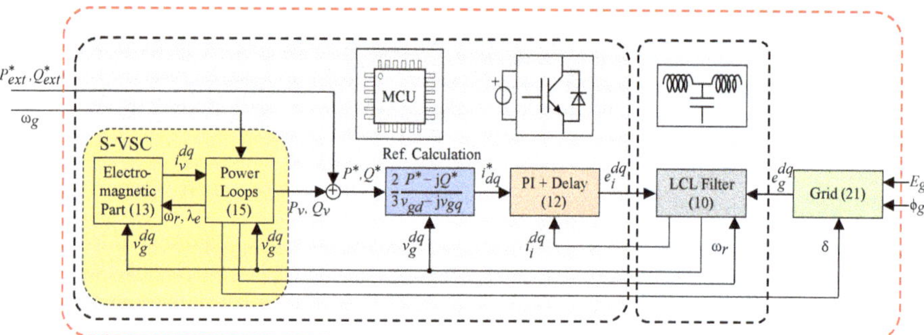

Figure 4. Block diagram of the modeled S-VSC [23] control with the physical system.

4.3. Dynamic Components Definition

4.3.1. LCL Filter

The inverter is connected to the grid by means of an LCL filter, as depicted in Figure 5. The filter is modeled in the (d,q) frame synchronous to the S-VSC virtual rotor position θ_r, rotating at ω_r.

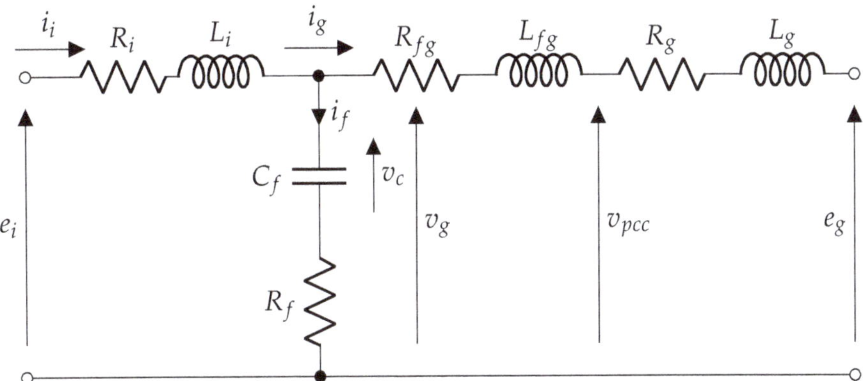

Figure 5. LCL model in the stationary (α,β) frame.

The equations modeling this component (in per unit) are as follows:

$$e_i^{dq} - v_g^{dq} = \frac{L_i}{\omega_b}\left(\frac{di_i^{dq}}{dt} + j\omega_r\omega_b i_i^{dq}\right) + R_i i_i^{dq}$$

$$v_g^{dq} - e_g^{dq} = \frac{L_{fg} + L_g}{\omega_b}\left(\frac{di_g^{dq}}{dt} + j\omega_r\omega_b i_g^{dq}\right) + (R_{fg} + R_g)i_g^{dq} \qquad (10)$$

$$\frac{C_f}{\omega_b}\frac{dv_c^{dq}}{dt} = i_i^{dq} - i_g^{dq} - jC_f\omega_r v_c^{dq}$$

This block has the following state variables X_{LCL}, inputs U_{LCL} and outputs Y_{LCL}:

$$X_{LCL} = \left[\Delta i_i^d, \Delta i_i^q, \Delta i_g^d, \Delta i_g^q, \Delta v_c^d, \Delta v_c^q\right]'_{6\times1}$$

$$U_{LCL} = \left[\Delta e_i^d, \Delta e_i^q, \Delta e_g^d, \Delta e_g^q, \Delta \omega_r\right]'_{5\times1} \qquad (11)$$

$$Y_{LCL} = \left[\Delta i_i^d, \Delta i_i^q, \Delta v_g^d, \Delta v_g^q, \Delta v_{pcc}^d, \Delta v_{pcc}^q\right]'_{6\times1}$$

The output v_{pcc} can be useful in the case of a multi-inverter plant where the voltage amplitude control at the plant point of common coupling (PCC) is required.

By linearizing (10), the necessary state-space matrices $A_{LCL}, B_{LCL}, C_{LCL}, D_{LCL}$ are obtained.

4.3.2. Current Controller

The current controller considered in this model has a sampling time of T_s and it is based on a proportional integral (PI) regulator implemented in the (d,q) synchronous reference. The proportional and integral gains of the PI regulator are k_p and k_i. The total delay $T_d = 1.5 \cdot T_s$ of the digital controller and the modulation is modeled with a first order Padé approximation. A higher order approximation can easily be considered by modifying this block. The controller complete block diagram is shown in Figure 6.

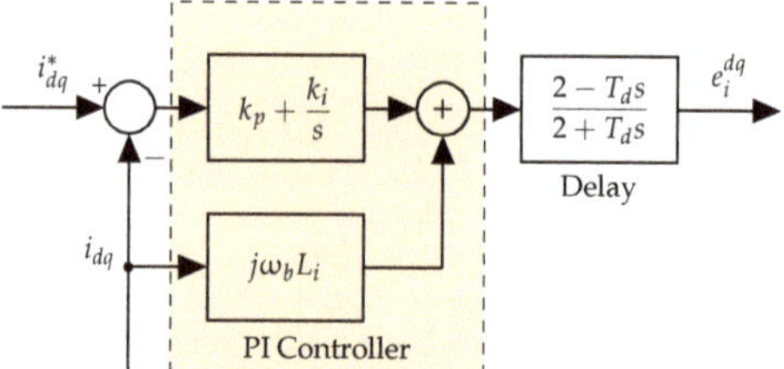

Figure 6. Block diagram of the PI regulator and equivalent delay (PI + delay) of the digital control and modulation.

The following state variables X_{Inv} are identified (the integrator Δx_i^{dq} and the delay model Δx_d^{dq}), as well as the inputs U_{Inv} and outputs Y_{Inv} of this block:

$$X_{Inv} = \left[\Delta x_i^d, \Delta x_i^q, \Delta x_d^d, \Delta x_d^q\right]'_{4\times 1}$$

$$U_{Inv} = \left[\Delta i_d^*, \Delta i_q^*, \Delta i_i^d, \Delta i_i^q\right]'_{4\times 1} \tag{12}$$

$$Y_{Inv} = \left[\Delta e_i^d, \Delta e_i^q\right]'_{2\times 1}$$

The state-space matrices of this block are $A_{Inv}, B_{Inv}, C_{Inv}, D_{Inv}$.

4.3.3. S-VSC Electromagnetic Equations

The electrical and magnetic equations of both virtual stator and rotor of the S-SVC [23] are (in per unit):

$$v_d = -R_s i_{vd} - \omega_r \lambda_q + \frac{1}{\omega_b}\frac{d\lambda_d}{dt}$$

$$v_q = -R_s i_{vq} + \omega_r \lambda_d + \frac{1}{\omega_b}\frac{d\lambda_q}{dt}$$

$$\tau_{rq0}\frac{d\lambda_{rq}}{dt} = -\lambda_{rq} - L_{rq} i_{vq} \tag{13}$$

$$i_{vd} = \frac{\lambda_e - \lambda_d}{L_s}$$

$$i_{vq} = \frac{\lambda_{rq} - \lambda_q}{L_s}$$

where λ_d, λ_q, λ_{rq} and λ_e are the virtual flux linkages of the machine (d-axis, q-axis, damper winding and excitation, respectively); i_{vd} and i_{vq} are the machine virtual currents; R_s and L_s are the virtual stator resistance and inductance; L_{rq} and τ_{rq0} are the virtual damper parameters, tuned as described in [38].

Therefore, the block state variables X_{Elt}, inputs U_{Elt} and outputs Y_{Elt} are defined as follows:

$$X_{Elt} = \left[\Delta\lambda_d, \Delta\lambda_q, \Delta\lambda_{rq}\right]'_{3\times 1}$$

$$U_{Elt} = \left[\Delta v_g^d, \Delta v_g^q, \Delta\omega_r, \Delta\lambda_e\right]'_{4\times 1} \tag{14}$$

$$Y_{Elt} = \left[\Delta i_v^d, \Delta i_v^q\right]'_{2\times 1}$$

The state-space matrices $A_{Elt}, B_{Elt}, C_{Elt}, D_{Elt}$ of this block are obtained from the linearization of (13).

4.3.4. S-VSC Power Loops Equations

The active and reactive power control equations of the S-SVC [23] are (in per unit):

$$
\begin{aligned}
-P_v &= 2H\frac{d\omega_r}{dt} \\[2mm]
\frac{d\delta}{dt} &= (\omega_r - \omega_g)\,\omega_b \\[2mm]
-k_e\frac{Q_v}{V_g} &= \frac{d\lambda_e}{dt}
\end{aligned}
\tag{15}
$$

where P_v and Q_v are the virtual active and reactive power of the S-VSC. The S-VSC is always operated at zero reference power in order to obtain better stability and damping [23]. H is the inertia constant of the machine, δ is the virtual load angle of the machine (q-axis to grid voltage vector). V_g is the peak value of the voltage across the filter capacitor C_f and k_e is the gain of the excitation control.

This block state variables X_{Power}, inputs U_{Power} and outputs Y_{Power} are as follows:

$$
\begin{aligned}
X_{Power} &= [\Delta\omega, \Delta\delta, \Delta\lambda_e]'_{3\times1} \\[2mm]
U_{Power} &= \left[\Delta v_g^d, \Delta v_g^q, \Delta i_v^d, \Delta i_v^q, \Delta\omega_g\right]'_{5\times1} \\[2mm]
Y_{Power} &= [\Delta P_v, \Delta Q_v, \Delta\omega_r, \Delta\delta, \Delta\lambda_e]'_{5\times1}
\end{aligned}
\tag{16}
$$

The state-space matrices A_{Power}, B_{Power}, C_{Power}, D_{Power} of the power loops block are again obtained by the linearization of (15).

4.4. Algebraic Components Definition

4.4.1. Power Reference Calculation

In this algebraic block the following calculation is performed to obtain the current references i_{dq}^* from the power references PQ^* and the grid voltage v_g^{dq}:

$$
i_d^* + ji_q^* = \frac{P^* - jQ^*}{v_{gd} - jv_{gq}}
\tag{17}
$$

where:

$$
\begin{aligned}
P^* &= P_{ext}^* + P_v \\[2mm]
Q^* &= Q_{ext}^* + Q_v
\end{aligned}
\tag{18}
$$

Are the sums of the external references and the S-VSC compensation power references.

Due to the division involved in (17), a dedicated block is justified to simplify the modeling process. There are no state variables, while the inputs U_{Ref} and outputs Y_{Ref} of the block are:

$$
\begin{aligned}
U_{Ref} &= \left[\Delta P_{ext}^*, \Delta Q_{ext}^*, \Delta P_v, \Delta Q_v, \Delta v_g^d, \Delta v_g^q\right]'_{6\times1} \\[2mm]
Y_{Ref} &= \left[\Delta i_d^*, \Delta i_q^*\right]'_{2\times1}
\end{aligned}
\tag{19}
$$

Being an algebraic block, only the D_{Ref} matrix is obtained.

4.4.2. Grid Perturbations

This algebraic block generates the grid voltage variations e_g^{dq} according to the external disturbances $\Delta\omega_g, \Delta E_g, \Delta\phi_g$, given as system inputs.

The grid voltage vector is defined as:

$$e_g^{dq} = E_g e^{-j\left(\frac{\pi}{2}-\delta\right)} \tag{20}$$

Any phase displacement $\Delta\phi_g$ in the grid results in the variation of the load angle δ. As seen in the previous S-VSC power loops section, the grid frequency $\Delta\omega_g$ variation leads to a load angle variation and it is, therefore, included in (15). The vector diagram of the grid and S-VSC is shown in Figure 7, as well as the angle variation due to a phase jump in the grid.

This block is defined by the following inputs U_{Grid} and outputs Y_{Grid}:

$$\begin{aligned}
U_{Grid} &= \left[\Delta\delta, \Delta E_g, \Delta\phi_g\right]'_{3\times1} \\
Y_{Grid} &= \left[\Delta e_g^d, \Delta e_g^q\right]'_{2\times1}
\end{aligned} \tag{21}$$

And from the matrix D_{Grid}, obtained by linearizing (20).

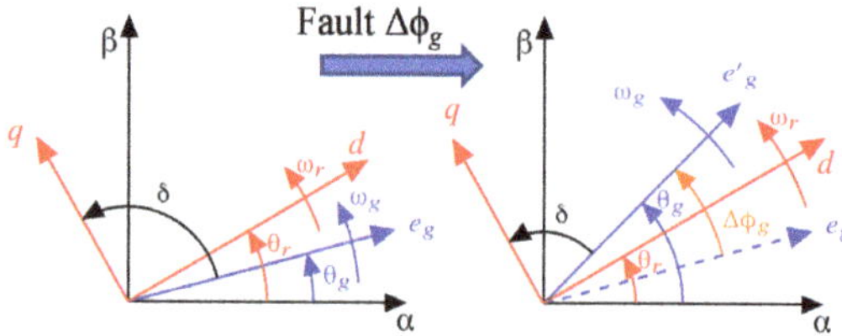

Figure 7. Vector diagram of the grid voltage and the (*d*,*q*) S-VSC rotating reference frames. The angle $\Delta\phi_g$ is defined as the load angle variation after a fault event.

4.5. Aggregated Model

Once all the dynamic and algebraic blocks have been defined, the aggregated model can be obtained. First the aggregated state variables X_s, inputs U and outputs Y are defined:

$$\begin{aligned}
X_s &= \left[X_{LCL}, X_{Inv}, X_{Elt}, X_{Power}\right]'_{16\times1} \\
U &= \left[U_{LCL}, U_{Inv}, U_{Elt}, U_{Power}, U_{Ref}, U_{Grid}\right]'_{27\times1} \\
Y &= \left[Y_{LCL}, Y_{Inv}, Y_{Elt}, Y_{Power}, Y_{Ref}, Y_{Grid}\right]'_{19\times1}
\end{aligned} \tag{22}$$

Then, the aggregated model matrices are:

$$
A_a = \begin{bmatrix} A_{LCL} & 0 & 0 & 0 \\ 0 & A_{Inv} & 0 & 0 \\ 0 & 0 & A_{Elt} & 0 \\ 0 & 0 & 0 & A_{Power} \end{bmatrix}_{16 \times 16}
$$

$$
B_a = \begin{bmatrix} B_{LCL} & 0 & 0 & 0 & \cdots & 0 \\ 0 & B_{Inv} & 0 & 0 & \cdots & 0 \\ 0 & 0 & B_{Elt} & 0 & \cdots & 0 \\ 0 & 0 & 0 & B_{Power} & \cdots & 0 \end{bmatrix}_{16 \times 27}
$$

$$
C_a = \begin{bmatrix} C_{LCL} & 0 & 0 & 0 \\ 0 & C_{Inv} & 0 & 0 \\ 0 & 0 & C_{Elt} & 0 \\ 0 & 0 & 0 & C_{Power} \\ \vdots & \vdots & \vdots & \vdots \\ 0 & 0 & 0 & 0 \end{bmatrix}_{19 \times 16} \tag{23}
$$

$$
D_a = \begin{bmatrix} D_{LCL} & 0 & 0 & 0 & 0 & 0 \\ 0 & D_{Inv} & 0 & 0 & 0 & 0 \\ 0 & 0 & D_{Elt} & 0 & 0 & 0 \\ 0 & 0 & 0 & D_{Power} & 0 & 0 \\ 0 & 0 & 0 & 0 & D_{Ref} & 0 \\ 0 & 0 & 0 & 0 & 0 & D_{Grid} \end{bmatrix}_{19 \times 29} .
$$

Note that the matrices B_a and C_a are extended with several null elements equal to the number of inputs (8 in total) and outputs (4 in total) of the two algebraic blocks. The detailed expressions of the matrices in (23) are available in the Appendix A.

4.6. Connection Matrices

Now the connection matrices can be obtained by linking the inputs U and the outputs Y of the aggregated system together with the global inputs U_s and outputs Y_s. The matrix T_{ss} is zero, since there is no direct feed-through from the system inputs to the system outputs. The detailed expressions of the connection matrices are available in the Appendix A.

$$
\begin{array}{cc} T_{uy\,27 \times 19} & T_{us\,27 \times 5} \\[6pt] T_{sy\,6 \times 19} & T_{ss\,6 \times 5} = 0_{6 \times 5} \end{array} \tag{24}
$$

4.7. Derivation of the Global State Space Model

Finally, by applying (8), the system matrices A_s, B_s, C_s, D_s are obtained. It must be noted that such model could have also been derived in a monolithic way, by studying the converter as a whole. However, by applying CCM, the model can be handled more easily and allows a quick and straightforward modification, in case some parts of the controller or of the physical system are changed. A practical and straightforward example of this advantage is the modification of the current feedback of the current controller. Usually, for grid-connected converters interfaced with LCL filters to the grid, two types of current feedbacks are adopted: grid side current feedback and converter side current feedback [39]. In the proposed example, the converter side current feedback was adopted. Therefore, the outputs of the LCL filter block Y_{LCL} features the converter side current. However, if the current feedback is shifted to the grid side current, only the Y_{LCL} vector and the relative matrices C_{LCL} and D_{LCL} must be modified. Each other block is not altered by this modification. Also the interconnection matrices are kept constant, being the two feedbacks equivalent from a signal routing point of view. A second example of the reconfiguration by adding new blocks is represented by the addition of an external droop controller. If a primary frequency regulation is required to the plant, then an extra proportional frequency controller is added, in order to generate the necessary active power references for the control. With the adopted CCM, the existing model is not modified, but an extra algebraic block is added before the reference calculation block. This extra block will receive as inputs the nominal and the actual grid frequency and generate the active droop power reference, which is then summed to the external power references.

5. Analysis and Validation

5.1. Analysis of the Derived State-Space Model

The poles λ of the system can be obtained numerically by solving:

$$det(A_s - \lambda I) = 0 \tag{25}$$

In Figure 8, the poles of the system (also listed in Table 1) are depicted in the complex plane. The parameters of the state-space model have been chosen according to the experimental setup and are listed in Table 2. The classical observations about the stability and damping of the system can be done. Moreover, the poles of the system can visually be distinguished as follows:

- The high frequency poles (1–4) are related to the LCL filter (in the range of 2 kHz);
- the equivalent poles of the digital control (5–8, 11, 12);
- the poles (9, 10) of the virtual stator of the S-VSC and the grid (50 Hz);
- the low frequency poles (13–16) of the electromechanical part of the S-VSC (excitation control, damper winding and swing equation) in the range of 1 Hz.

To clearly define which phenomena are related to the single poles, the analysis of the participation factors [14] can be performed. The participation factor p_{ik} of the k-th state to the i-th mode is defined as:

$$p_{ik} = \Phi_{ik} \cdot \Psi_{ki} \tag{26}$$

where Φ_{ik} is the k-th value of the right eigenvector Φ_i and Ψ_{ki} is the k-th value of the left eigenvector Ψ_i.

From (26), the states and the modes of the system can be correlated as the results in Table 1. In this way, the poles can be associated with the physical quantities that influence them.

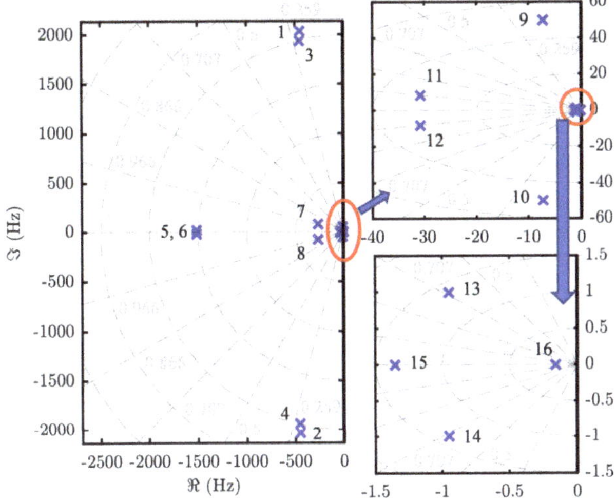

Figure 8. Pole map of the S-SVC. The poles are numbered according to Table 1.

Table 1. Poles and participation factors. Only significant factors are displayed (larger than 0.1).

Poles	1–2	3–4	5–6	7–8	9–10	11–12	13–14	15	16
f_0 (Hz)	2084.58	1988.07	1513.86	272.92	50.34	31.85	1.38	1.35	0.16
τ (ms)	0.35	0.35	0.11	0.61	21.87	5.17	167.39	117.67	999.67
ζ	0.216	0.229	~1	0.959	0.145	0.966	0.691	–	–
i_{id}	0.2065	0.2079	0.1532	0.2639	–	–	–	–	–
i_{iq}	0.2065	0.2079	0.1532	0.2609	–	–	–	–	–
i_{gd}	–	–	–	–	0.2740	–	–	–	–
i_{gq}	–	–	–	–	0.2810	–	–	–	–
v_{gd}	0.2471	0.2475	–	–	–	–	–	–	–
v_{gq}	0.2471	0.2474	–	–	–	–	–	–	–
x_{id}	–	–	–	–	–	0.5750	–	–	–
x_{iq}	–	–	–	–	–	0.5759	–	–	–
x_{pd}	–	–	0.6859	0.1876	–	–	–	–	–
x_{pq}	–	–	0.6863	0.1875	–	–	–	–	–
λ_d	–	–	–	0.4200	0.2282	–	–	–	–
λ_q	–	–	–	0.4178	0.2199	–	–	–	–
λ_{rq}	–	–	–	–	–	–	2.2608	5.5357	–
ω_r	–	–	–	–	–	–	1.6308	−2.2624	–
δ	–	–	–	–	–	–	1.6362	−2.2699	–
λ_e	–	–	–	–	–	–	–	–	1.0002

Table 2. Inverter and S-VSC parameters for the state-space model and of the experimental setup.

Parameter	Value	Parameter	Value
V_b	$230\sqrt{2}$ V	S_b	15 kVA
L_f	2 mH	C_f	5 μF
L_{fg}	1 mH	L_g	3 mH
k_p	3.77 Ω	k_i	710.6 Ω/s
L_s	0.1 pu	R_s	0.02 pu
H	4 s	k_e	0.22
L_{rq}	1.048 pu	τ_{rq0}	0.278 s
$f_s = f_{sw}$	10 kHz	V_{DC}	700 V

From the point of view of a multi-timescale analysis, the poles whose time constants are too small or too large can be excluded to simplify the model. For example, in the considered case study, the model can be simplified to analyze the low frequency behavior of the system (e.g., integration in a wider network to study the inertial behavior of the system during a frequency drop), by neglecting the high frequency poles related to the LCL filter physical parameters and from the current controller, which can be approximated with a direct feed-through with no dynamics, being orders of magnitude faster than the electromechanical dynamics of the S-VSC. The blocks and their interconnections can be rearranged to obtain a simplified model. This simplified model is useful to study the low frequency interaction within a more complex network.

Moreover, the state-space model allows the study of the behavior of the system under a change of parameters. A practical example is the variation of the short circuit ratio (SCR) in the connection to the grid, representing the inverse of the magnitude of the grid impedance in per unit. This analysis is especially important for wind power applications connected to very weak grids. In fact, due to the often isolated geographical positions, long cables or overhead lines connections are necessary and SCR ranging down to 1.5 or even less have been reported [40].

In Figure 9 the SCR of the system is swept from 10 to 1.5. The S-SVC damper and excitation control parameters are tuned accordingly. As it can be seen from the pole map, the poles relative to the LCL filter (1–4) are altered in their frequency, as the grid side inductance is modified. The S-VSC stator poles (9, 10) are also affected, being the stator in an equivalent series connection with the grid inductance. The effect is a lower damping of such poles that can be however easily compensated for by increasing the virtual stator resistance R_s. Finally, the low frequency mechanical poles (13–15) do not lose damping, thanks to the tuning algorithm described in [38], but only change their natural frequency. The excitation control pole (16) is not modified, as it is tuned to always obtain the desired time constant.

As previously mentioned, the state-space model can be used for a wide range of analyses. Analyses of interest are, for example, the pole map for different operating points (i.e., operation under load) and the pole variation due to the influence of parametric uncertainty of the physical components (i.e., grid inductance, filter parameters) on the tuning of both the controllers and the VSG.

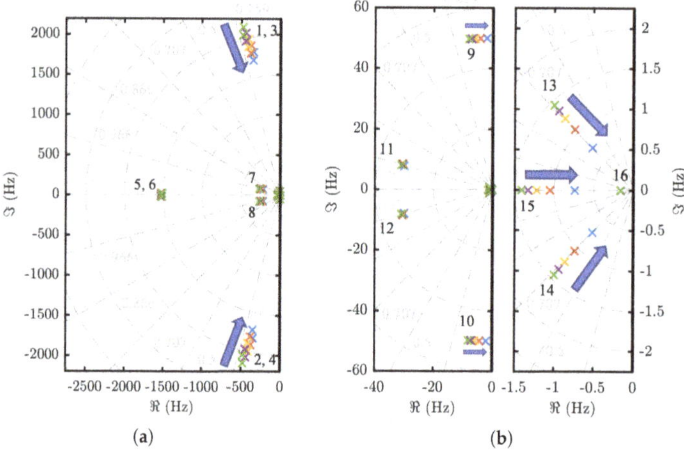

Figure 9. Pole map of the S-SVC when the SCR changes from 10 to 1.5. The arrows indicate a decrease in the SCR. The poles are numbered according to Table 1. From left to right: (**a**) Complete pole map; (**b**) Magnifications of the lower frequency poles.

5.2. Experimental Validation

The obtained state-space model can then be verified by a comparison with either a simulation of the complete system (e.g., Simulink, PLECS, PSCAD and other simulation tools can be used) or experimentally.

In this case, the S-VSC has been implemented on a dSPACE DS 1007, controlling a 15 kVA three-phase inverter connected to a grid emulator, as shown in Figure 10. The control runs at $f_s = 10$ kHz, which also corresponds to the switching frequency f_{sw} of the converter.

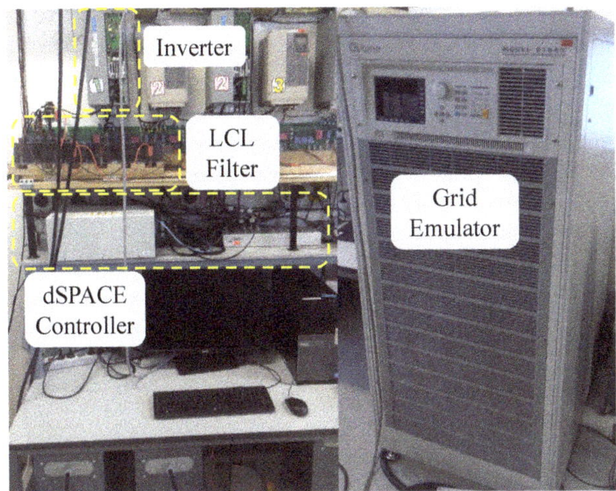

Figure 10. Experimental setup where the controller is based on dSPACE.

The parameters of both the experimental setup and the S-VSC are listed in Table 2.

The following four tests have been performed to validate the state-space model using all the five available inputs ΔP_{ext}^*, ΔQ_{ext}^*, $\Delta \omega_g$, ΔE_g and $\Delta \phi_g$:

1. Step active power reference variation ΔP_{ext}^*;
2. Step reactive power reference variation ΔQ_{ext}^*;
3. Step drop of Grid frequency $\Delta \omega_g$;
4. Grid voltage drop ΔE_g with phase jump $\Delta \phi_g$.

In the first test (Figure 11) the active power fed to the grid and the frequency of the S-VSC are compared with the outputs of the state-space model. The active power reference rises from 0.2 pu to 0.3 pu at $t = 0.2$ s. This choice allows a more general test of the state-space model. In fact, as it can be seen in (A5), a large part of the matrix D_s depends on the initial active and reactive power references P_0^* and Q_0^*. Therefore, a non zero initial power operating point has been chosen, better to verify this part of the model. As it can be seen from Figure 11 both the S-VSC virtual speed ω_r and the injected active power are similar to the results from the state-space model. The state space-model, as it has been described, does not take into account the non-ideal behavior of the measurement process (e.g., noise, delays), which on the other hand may affect the experimental setup.

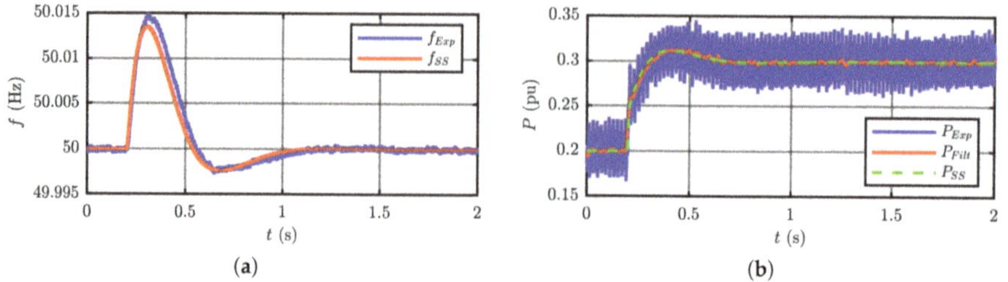

Figure 11. Test 1: Step in active power reference ΔP^*_{ext} from 0.2 pu to 0.3 pu. From left to right: (**a**) S-VSC virtual rotor speed (Hz) from experimental test (Exp) and state-space model (SS); (**b**) Active power injected from the inverter (pu) from experimental test (Exp), filtered active power (Filt) and output of the state-space model (SS).

In the second test (Figure 12), the reactive power injected by the inverter is compared, when the reactive power reference rises from 0.1 pu to 0.2 pu. As with Test 1, the initial non-zero operating point has been chosen better to test the model. Also in this case, the obtained model is accurate enough to simulate the step transient in the reactive power reference. Again, the state-space model is not including the non-ideal behavior of the experimental setup. This non-ideal behavior can be neglected when dealing with longer timescales, such as in Test 2. In fact, the delays and the noise act with much shorter time constants, which do not affect the behavior during these transients (operating in the time scale of seconds).

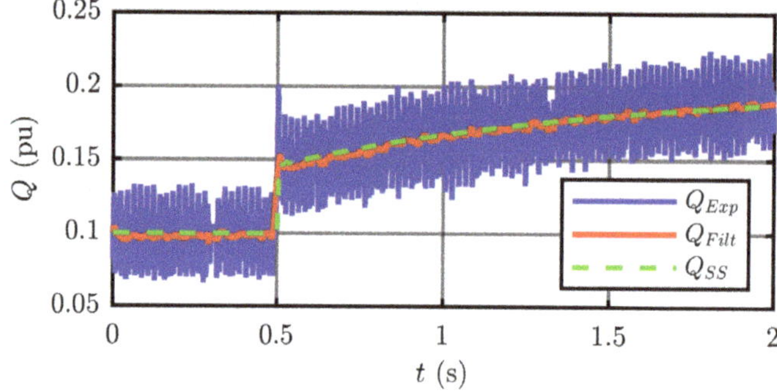

Figure 12. Test 2: Step in reactive power reference ΔQ^*_{ext} from 0.1 pu to 0.2 pu. Reactive power injected from the inverter (pu) from experimental test (exp), filtered reactive power (Filt) and output of the state-space model (SS).

Test 3 (Figure 13) deals with the input relative to the grid frequency variation $\Delta \omega_g$. A step frequency variation of -0.2 Hz has been applied by the grid emulator. The S-VSC tries to compensate the frequency drop by injecting active power into the grid, providing a virtual inertial effect. The state-space model well predicts both the S-VSC speed ω_r profile (Figure 13a) and the amount of active power injected (Figure 13b). The modeling of the frequency variations, as obtained here, can be useful in system level studies, when analyzing frequency regulation and inertial support. This kind of analysis also helps studying the inertial effect resulting from a different tuning of the S-VSC (the key parameters are in this case the inertia constant H and the virtual damper parameters τ_{rq0} and L_{rq}, which are directly related to the electromechanical damping ζ of the virtual machine [38]).

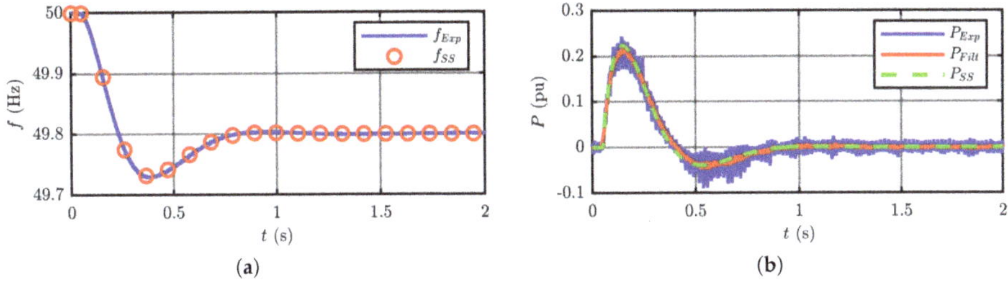

Figure 13. Test 3: Grid frequency step drop $\Delta\omega_g$ from 50 Hz to 49.8 Hz. From left to right: (**a**) S-VSC virtual rotor speed ω_r (Hz) from experimental test (Exp) and state-space model (SS); (**b**) Active power injected into the grid (pu) from experimental test (Exp), filtered active power (Filt) and output of the state-space model (SS).

Finally, Test 4 (Figure 14) validates the behavior of the system in case of a realistic grid fault. During the fault, the voltage of the grid drops to 95% of its nominal value and a phase jump of $-5°$ is applied. For such tests, the quantities of interest are the reactive power injected into the grid (Figure 14a) to provide reactive support and the virtual excitation flux of the S-VSC λ_e (Figure 14b) to evaluate the time constant of the excitation control. As expected, the state-space model is well predicting both the shape and the amplitude of such quantities, proving itself as a good analysis tool to correctly tune the S-VSC to guarantee the desired reactive response during grid faults.

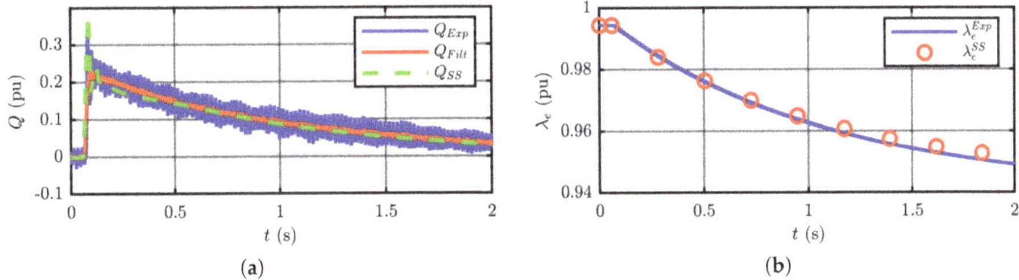

Figure 14. Test 4: Grid voltage drop $\Delta E_g = -5\%$ with phase jump $\Delta\phi_g = -5°$. From left to right: (**a**) Reactive power injected into the grid (pu) from experimental test (Exp), filtered reactive power (Filt) and output of the state-space model (SS); (**b**) S-VSC virtual excitation flux λ_e from the experimental test (Exp) and state-space model (SS).

6. Conclusions

In this paper, a general step-by-step procedure to easily obtain the state-space model of a grid-connected converter equipped with advanced cascaded controllers using CCM is presented. This mathematical method is suitable to model cascaded controllers with interconnected signals thanks to its modular approach. Moreover, it is easily possible to modify or reconfigure the system, without remodeling it.

This approach can be used to derive state-space models of grid-tied inverters equipped with advanced cascaded controllers or emerging control techniques, such as VSGs in order to study and simulate the stability of a power system with a higher integration of renewable energy sources. In addition, state-space models are favorable in case of multi-timescale analysis of the system: the eigenvalues of the system can be obtained, as well as their correlation to the system state variables. Therefore, simplified models can be obtained, neglecting high or low frequency phenomena which are out of the scope of the analysis.

The single models obtained following this procedure can represent a useful tool for both power electronics designers and grid operators. The first can obtain preliminary information on the weak spots of the system they are designing and test their behavior under different operating conditions. From the grid side, such models would be provided by the manufacturers with different level of details [3,4] to be integrated in a more system-level analysis.

This step-by-step approach has been verified with a practical example of the modeling of a VSG model (S-VSC), which is then validated experimentally on a 15 kVA grid-connected inverter.

Author Contributions: Conceptualization, F.M., E.C., R.B., T.D. and F.B.; methodology, F.M. and E.C.; software, F.M.; validation, F.M.; formal analysis, F.M. and E.C.; investigation, F.M. and E.C.; resources, S.M., R.B. and F.B.; data curation, F.M.; writing–original draft preparation, F.M.; writing–review and editing, F.M., R.B., T.D. and F.B.; visualization, F.M., S.M., R.B., T.D. and F.B.; supervision, R.B., T.D. and F.B.; project administration, S.M., R.B., T.D. and F.B. All authors have read and agreed to the published version of the manuscript.

Funding: This research received no external funding.

Conflicts of Interest: The authors declare no conflict of interest.

Abbreviations

The following abbreviations are used in this manuscript:

CCM	Component Connection Method
VSG	Virtual Synchronous Generator
HVDC	High Voltage Direct Current
S-VSC	Simplified Virtual Synchronous Compensator
PLL	Phase Locked Loop
PI	Proportional Integral

Appendix A

The subscript 0 indicates the value around which the linearization is performed.

LCL filter matrices:

$$A_{LCL} = \begin{bmatrix} -\omega_b\dfrac{R_i+R_f}{L_i} & \omega_{r0}\omega_b & \omega_b\dfrac{R_f}{L_i} & 0 & -\dfrac{\omega_b}{L_i} & 0 \\[2ex] -\omega_{r0}\omega_b & -\omega_b\dfrac{R_i+R_f}{L_i} & 0 & \omega_b\dfrac{R_f}{L_i} & 0 & -\dfrac{\omega_b}{L_i} \\[2ex] \omega_b\dfrac{R_f}{L_{fg}+L_g} & 0 & -\omega_b\dfrac{R_{fg}+R_g+R_f}{L_{fg}+L_g} & \omega_{r0}\omega_b & \dfrac{\omega_b}{L_{fg}+L_g} & 0 \\[2ex] 0 & \omega_b\dfrac{R_f}{L_{fg}+L_g} & -\omega_{r0}\omega_b & -\omega_b\dfrac{R_{fg}+R_g+R_f}{L_{fg}+L_g} & 0 & \dfrac{\omega_b}{L_{fg}+L_g} \\[2ex] \dfrac{\omega_b}{C_f} & 0 & -\dfrac{\omega_b}{C_f} & 0 & 0 & \omega_{r0}\omega_b \\[2ex] 0 & \dfrac{\omega_b}{C_f} & 0 & -\dfrac{\omega_b}{C_f} & -\omega_{r0}\omega_b & 0 \end{bmatrix}$$

$$B_{LCL} = \begin{bmatrix} \dfrac{\omega_b}{L_i} & 0 & 0 & 0 & \omega_b I_{i0}^{q} \\[2ex] 0 & \dfrac{\omega_b}{L_i} & 0 & 0 & -\omega_b I_{i0}^{d} \\[2ex] 0 & 0 & -\dfrac{\omega_b}{L_{fg}+L_g} & 0 & \omega_b I_{g0}^{q} \\[2ex] 0 & 0 & 0 & -\dfrac{\omega_b}{L_{fg}+L_g} & -\omega_b I_{g0}^{d} \\[2ex] 0 & 0 & 0 & 0 & \omega_b V_{c0}^{q} \\[2ex] 0 & 0 & 0 & 0 & -\omega_b V_{c0}^{d} \end{bmatrix} \tag{A1}$$

$$C_{LCL} = \begin{bmatrix} 1 & 0 & 0 & 0 & 0 & 0 \\[1ex] 0 & 1 & 0 & 0 & 0 & 0 \\[1ex] R_f & 0 & -R_f & 0 & 1 & 0 \\[1ex] 0 & R_f & 0 & -R_f & 0 & 1 \\[1ex] \dfrac{L_g R_f}{L_{fg}+L_g} & 0 & \dfrac{R_g L_{fg}+R_{fg}L_g-L_g R_f}{L_{fg}+L_g} & 0 & \dfrac{L_g}{L_{fg}+L_g} & 0 \\[2ex] 0 & \dfrac{L_g R_f}{L_{fg}+L_g} & 0 & \dfrac{R_g L_{fg}+R_{fg}L_g-L_g R_f}{L_{fg}+L_g} & 0 & \dfrac{L_g}{L_{fg}+L_g} \end{bmatrix}$$

$$D_{LCL} = \begin{bmatrix} 0 & 0 & 0 & 0 & 0 \\[1ex] 0 & 0 & 0 & 0 & 0 \\[1ex] 0 & 0 & 0 & 0 & 0 \\[1ex] 0 & 0 & 0 & 0 & 0 \\[1ex] 0 & 0 & \dfrac{L_{fg}}{L_{fg}+L_g} & 0 & 0 \\[2ex] 0 & 0 & 0 & \dfrac{L_{fg}}{L_{fg}+L_g} & 0 \end{bmatrix}$$

Current controller matrices:

$$A_{Inv} = \begin{bmatrix} 0 & 0 & 0 & 0 \\ 0 & 0 & 0 & 0 \\ \dfrac{4}{T_d} & 0 & -\dfrac{2}{T_d} & 0 \\ 0 & \dfrac{4}{T_d} & 0 & -\dfrac{2}{T_d} \end{bmatrix}$$

$$B_{Inv} = \begin{bmatrix} k_i & 0 & -k_i & 0 \\ 0 & k_i & 0 & -k_i \\ k_p\dfrac{4}{T_d} & 0 & -k_p\dfrac{4}{T_d} & \omega_{r0}L_i\dfrac{4}{T_d} \\ 0 & k_p\dfrac{4}{T_d} & -\omega_{r0}L_i\dfrac{4}{T_d} & -k_p\dfrac{4}{T_d} \end{bmatrix} \tag{A2}$$

$$C_{Inv} = \begin{bmatrix} -1 & 0 & 1 & 0 \\ 0 & -1 & 0 & 1 \end{bmatrix}$$

$$D_{Inv} = \begin{bmatrix} -k_p & 0 & k_p & -\omega_{r0}L_i \\ 0 & -k_p & \omega_{r0}L_i & k_p \end{bmatrix}$$

S-VSC electromagnetic part matrices:

$$A_{Elt} = \begin{bmatrix} -\omega_b\dfrac{R_v}{L_d''} & \omega_{r0}\omega_b & 0 \\ -\omega_{r0}\omega_b & -\omega_b\dfrac{R_v}{L_q''} & \omega_b\dfrac{R_v}{L_q''} \\ 0 & \dfrac{L_{rq}}{\tau_{rq0}L_q''} & -\dfrac{1+L_{rq}/L_q''}{\tau_{rq0}} \end{bmatrix}$$

$$B_{Elt} = \begin{bmatrix} \omega_b & 0 & \omega_b\Lambda_{q0} & \omega_b\dfrac{R_v}{L_d''} \\ 0 & \omega_b & -\omega_b\Lambda_{d0} & 0 \\ 0 & 0 & 0 & 0 \end{bmatrix} \tag{A3}$$

$$C_{Elt} = \begin{bmatrix} -\dfrac{1}{L_d''} & 0 & 0 \\ 0 & -\dfrac{1}{L_q''} & \dfrac{1}{L_q''} \end{bmatrix}$$

$$D_{Elt} = \begin{bmatrix} 0 & 0 & 0 & \dfrac{1}{L_d''} \\ 0 & 0 & 0 & 0 \end{bmatrix}$$

S-VSC power loops matrices:

$$A_{Power} = \begin{bmatrix} 0 & 0 & 0 \\ \omega_b & 0 & 0 \\ 0 & 0 & 0 \end{bmatrix}$$

$$B_{Power} = \begin{bmatrix} 0 & 0 & -\dfrac{I_{vd0}}{2H} & -\dfrac{I_{vq0}}{2H} & -\dfrac{V_{gd0}}{2H} & -\dfrac{V_{gq0}}{2H} & 0 \\ 0 & 0 & 0 & 0 & 0 & 0 & -\omega_b \\ 0 & 0 & k_e\dfrac{I_{vq0}}{V_{g0}} & -k_e\dfrac{I_{vd0}}{V_{g0}} & -k_e\dfrac{V_{gq0}}{V_{g0}} & k_e\dfrac{V_{gd0}}{V_{g0}} & 0 \end{bmatrix}$$

$$C_{Power} = \begin{bmatrix} 0 & 0 & 0 \\ 0 & 0 & 0 \\ 1 & 0 & 0 \\ 0 & 1 & 0 \\ 0 & 0 & 1 \end{bmatrix}$$

$$D_{Power} = \begin{bmatrix} 0 & 0 & I_{vd0} & I_{vq0} & V_{gd0} & V_{gq0} & 0 \\ 0 & 0 & -I_{vq0} & I_{vd0} & V_{gq0} & -V_{gd0} & 0 \\ 0 & 0 & 0 & 0 & 0 & 0 & 0 \\ 0 & 0 & 0 & 0 & 0 & 0 & 0 \\ 0 & 0 & 0 & 0 & 0 & 0 & 0 \end{bmatrix}$$

(A4)

Power to current reference calculation:

$$D_{Ref} =$$

$$\begin{bmatrix} \dfrac{v_{gd0}}{V_{g0}^2} & \dfrac{v_{gq0}}{V_{g0}^2} & \dfrac{v_{gd0}}{V_{g0}^2} & \dfrac{v_{gq0}}{V_{g0}^2} & \dfrac{v_{gd0}}{V_{g0}^2} & \dfrac{v_{gq0}}{V_{g0}^2} & \dfrac{P_0^*(v_{gq0}^2 - v_{gd0}^2) - 2Q_0^* v_{gd0} v_{gq0}}{V_{g0}^4} & \dfrac{Q_0^*(v_{gd0}^2 - v_{gq0}^2) - 2P_0^* v_{gd0} v_{gq0}}{V_{g0}^4} \\ \dfrac{v_{gq0}}{V_{g0}^2} & -\dfrac{v_{gd0}}{V_{g0}^2} & \dfrac{v_{gq0}}{V_{g0}^2} & -\dfrac{v_{gd0}}{V_{g0}^2} & \dfrac{v_{gq0}}{V_{g0}^2} & -\dfrac{v_{gd0}}{V_{g0}^2} & \dfrac{2P_0^* v_{gd0} v_{gq0} + Q_0^*(v_{gq0}^2 - v_{gd0}^2)}{V_{g0}^4} & \dfrac{P_0^*(v_{gd0}^2 - v_{gq0}^2) + 2Q_0^* v_{gd0} v_{gq0}}{V_{g0}^4} \end{bmatrix}$$

(A5)

Grid perturbation matrix:

$$D_{Grid} = \begin{bmatrix} e_{gq0} & \dfrac{e_{gd0}}{E_{g0}} & -e_{gq0} \\ -e_{gd0} & \dfrac{e_{gq0}}{E_{g0}} & e_{gd0} \end{bmatrix}$$

(A6)

Connection matrices. Only the non-zero elements are given:

$$T_{uy\,27\times19} : \begin{array}{lllll} T_{uy}(1,7) = 1 & T_{uy}(2,8) = 1 & T_{uy}(3,18) = 1 & T_{uy}(4,19) = 1 & T_{uy}(5,13) = 1 \\ T_{uy}(6,16) = 1 & T_{uy}(7,17) = 1 & T_{uy}(8,1) = 1 & T_{uy}(9,2) = 1 & T_{uy}(10,3) = 1 \\ T_{uy}(11,4) = 1 & T_{uy}(12,13) = 1 & T_{uy}(13,15) = 1 & T_{uy}(14,3) = 1 & T_{uy}(15,4) = 1 \\ T_{uy}(16,9) = 1 & T_{uy}(17,10) = 1 & T_{uy}(21,11) = 1 & T_{uy}(22,12) = 1 & T_{uy}(23,3) = 1 \\ T_{uy}(24,4) = 1 & T_{uy}(25,14) = 1 \end{array}$$

(A7)

$$T_{us\,27\times5} : \quad T_{us}(18,3) = 1 \quad T_{us}(19,1) = 1 \quad T_{us}(20,2) = 1 \quad T_{us}(26,4) = 1 \quad T_{us}(27,5) = 1$$

$$T_{sy\,6\times19} : \begin{array}{llll} T_{sy}(1,1) = \dfrac{3}{2}V_{g0}^d & T_{sy}(1,3) = \dfrac{3}{2}I_{g0}^d & T_{sy}(1,2) = \dfrac{3}{2}V_{g0}^q & T_{sy}(1,4) = \dfrac{3}{2}I_{g0}^q \\ T_{sy}(2,2) = \dfrac{3}{2}V_{g0}^q & T_{sy}(2,4) = \dfrac{3}{2}I_{g0}^d & T_{sy}(2,1) = -\dfrac{3}{2}V_{g0}^d & T_{sy}(2,3) = -\dfrac{3}{2}I_{g0}^d \\ T_{sy}(3,13) = 1 & T_{sy}(4,14) = 1 & T_{sy}(5,3) = 1 & T_{sy}(6,4) = 1 \end{array}$$

References

1. Li, C. Unstable Operation of Photovoltaic Inverter from Field Experiences. *IEEE Trans. Power Deliv.* **2018**, *33*, 1013–1015. [CrossRef]
2. Mollerstedt, E.; Bernhardsson, B. Out of control because of harmonics-an analysis of the harmonic response of an inverter locomotive. *IEEE Control Syst. Mag.* **2000**, *20*, 70–81. [CrossRef]

3.   Terna Grid Code. *Allegato A17—CENTRALI EOLICHE—Condizioni Generali di Connessione Alle Reti AT Sistemi di Protezione Regolazione e Controllo*; Terna Grid Code Attachment A17; TERNA: Roma, Italy, 2018.

4.   Terna Grid Code. *Allegato A68—CENTRALI FOTOVOLTAICHE—Condizioni Generali di Connessione Alle Reti AT Sistemi di Protezione Regolazione e Controllo*; Terna Grid Code—Attachment A68; TERNA: Roma, Italy, 2018.

5.   Wang, X.; Blaabjerg, F. Harmonic Stability in Power Electronic-Based Power Systems: Concept, Modeling, and Analysis. *IEEE Trans. Smart Grid* **2019**, *10*, 2858–2870. [CrossRef]

6.   Sun, J. Small-Signal Methods for AC Distributed Power Systems—A Review. *IEEE Trans. Power Electron.* **2009**, *24*, 2545–2554. [CrossRef]

7.   Sun, J. Impedance-Based Stability Criterion for Grid-Connected Inverters. *IEEE Trans. Power Electron.* **2011**, *26*, 3075–3078. [CrossRef]

8.   Wang, X.; Blaabjerg, F.; Wu, W. Modeling and Analysis of Harmonic Stability in an AC Power-Electronics-Based Power System. *IEEE Trans. Power Electron.* **2014**, *29*, 6421–6432. [CrossRef]

9.   Wang, X.; Harnefors, L.; Blaabjerg, F. Unified Impedance Model of Grid-Connected Voltage-Source Converters. *IEEE Trans. Power Electron.* **2018**, *33*, 1775–1787. [CrossRef]

10.   Wen, B.; Boroyevich, D.; Burgos, R.; Mattavelli, P.; Shen, Z. Analysis of D-Q Small-Signal Impedance of Grid-Tied Inverters. *IEEE Trans. Power Electron.* **2016**, *31*, 675–687. [CrossRef]

11.   Liu, H.; Xie, X.; Liu, W. An Oscillatory Stability Criterion Based on the Unified dq-Frame Impedance Network Model for Power Systems With High-Penetration Renewables. *IEEE Trans. Power Syst.* **2018**, *33*, 3472–3485. [CrossRef]

12.   Francis, G.; Burgos, R.; Boroyevich, D.; Wang, F.; Karimi, K. An algorithm and implementation system for measuring impedance in the D-Q domain. In Proceedings of the 2011 IEEE Energy Conversion Congress and Exposition, Phoenix, AZ, USA, 17–22 September 2011; pp. 3221–3228. [CrossRef]

13.   Belkhayat, M. Stability Criteria for AC Power Systems with Regulated Loads. Ph.D. Thesis, Purdue University, West Lafayette, IN, USA, 1997.

14.   Kundur, P. *Power System Stability and Control*; McGraw-Hill Education: New York, NY, USA, 1994.

15.   Fang, J.; Li, H.; Tang, Y.; Blaabjerg, F. On the Inertia of Future More-Electronics Power Systems. *IEEE J. Emerg. Sel. Top. Power Electron.* **2019**, *7*, 2130–2146. [CrossRef]

16.   Beck, H.; Hesse, R. Virtual synchronous machine. In Proceedings of the 2007 9th International Conference on Electrical Power Quality and Utilisation, Barcelona, Spain, 9–11 October 2007; pp. 1–6. [CrossRef]

17.   Zhong, Q.; Weiss, G. Synchronverters: Inverters That Mimic Synchronous Generators. *IEEE Trans. Ind. Electron.* **2011**, *58*, 1259–1267. [CrossRef]

18.   D'Arco, S.; Suul, J.A.; Fosso, O.B. A Virtual Synchronous Machine implementation for distributed control of power converters in SmartGrids. *Electr. Power Syst. Res.* **2015**, *122*, 180–197. [CrossRef]

19.   Driesen, J.; Visscher, K. Virtual synchronous generators. In Proceedings of the 2008 IEEE Power and Energy Society General Meeting—Conversion and Delivery of Electrical Energy in the 21st Century, Pittsburgh, PA, USA, 20–24 July 2008; pp. 1–3. [CrossRef]

20.   Van Wesenbeeck, M.P.N.; De Haan, S.W.H.; Varela, P.; Visscher, K. Grid tied converter with virtual kinetic storage. In Proceedings of the 2009 IEEE Bucharest PowerTech, Bucharest, Romania, 28 June–2 July 2009; pp. 1–7. [CrossRef]

21.   Bevrani, H.; Ise, T.; Miura, Y. Virtual synchronous generators: A survey and new perspectives. *Int. J. Electr. Power Energy Syst.* **2014**, *54*, 244–254. [CrossRef]

22.   Rodríguez, P.; Citro, C.; Candela, J.I.; Rocabert, J.; Luna, A. Flexible Grid Connection and Islanding of SPC-Based PV Power Converters. *IEEE Trans. Ind. Appl.* **2018**, *54*, 2690–2702. [CrossRef]

23.   Mandrile, F.; Carpaneto, E.; Bojoi, R. Grid-Tied Inverter with Simplified Virtual Synchronous Compensator for Grid Services and Grid Support. In Proceedings of the 2019 IEEE Energy Conversion Congress and Exposition (ECCE), Baltimore, MD, USA, 29 September–3 October 2019; IEEE: Baltimore, MD, USA, 2019; pp. 4317–4323. [CrossRef]

24.   Coelho, E.A.A.; Cortizo, P.C.; Garcia, P.F.D. Small-signal stability for parallel-connected inverters in stand-alone AC supply systems. *IEEE Trans. Ind. Appl.* **2002**, *38*, 533–542. [CrossRef]

25.   Agorreta, J.L.; Borrega, M.; López, J.; Marroyo, L. Modeling and Control of N-Paralleled Grid-Connected Inverters With LCL Filter Coupled Due to Grid Impedance in PV Plants. *IEEE Trans. Power Electron.* **2011**, *26*, 770–785. [CrossRef]

26. Bottrell, N.; Prodanovic, M.; Green, T.C. Dynamic Stability of a Microgrid With an Active Load. *IEEE Trans. Power Electron.* **2013**, *28*, 5107–5119. [CrossRef]

27. Uudrill, J.M. Dynamic Stability Calculations for an Arbitrary Number of Interconnected Synchronous Machines. *IEEE Trans. Power Appar. Syst.* **1968**, *PAS-87*, 835–844. [CrossRef]

28. Gaba, G.; Lefebvre, S.; Mukhedkar, D. Comparative analysis and study of the dynamic stability of AC/DC systems. *IEEE Trans. Power Syst.* **1988**, *3*, 978–985. [CrossRef]

29. Arabi, S.; Rogers, G.; Wong, D.; Kundur, P.; Lauby, M. Small signal stability program analysis of SVC and HVDC in AC power systems. *IEEE Trans. Power Syst.* **1991**, *6*, 1147–1153. [CrossRef]

30. Hou, P.; Ebrahimzadeh, E.; Wang, X.; Blaabjerg, F.; Fang, J.; Wang, Y. Harmonic stability analysis of offshore wind farm with component connection method. In Proceedings of the IECON 2017—43rd Annual Conference of the IEEE Industrial Electronics Society, Beijing, China, 29 October–1 November 2017; IEEE: Beijing, China, 2017; pp. 4926–4932. [CrossRef]

31. Wang, Y. Stability Assessment of Inverter-Fed Power Systems. Ph.D. Thesis, Aalborg University, Aalborg, Denmark, 2017.

32. Rosso, R.; Engelken, S.; Liserre, M. Analysis of the Behavior of Synchronverters Operating in Parallel by Means of Component Connection Method (CCM). In Proceedings of the 2018 IEEE Energy Conversion Congress and Exposition (ECCE), Portland, OR, USA, 23–27 September 2018; IEEE: Portland, OR, USA, 2018; pp. 2228–2235. [CrossRef]

33. Rosso, R.; Engelken, S.; Liserre, M. Analysis of the Parallel Operation Between Synchronverters and PLL-Based Converters. In Proceedings of the 2019 IEEE Energy Conversion Congress and Exposition (ECCE), Baltimore, MD, USA, 29 September–3 October 2019; IEEE: Baltimore, MD, USA, 2019; pp. 2583–2590. [CrossRef]

34. Rosso, R.; Engelken, S.; Liserre, M. Robust Stability Investigation of the Interactions Among Grid-Forming and Grid-Following Converters. *IEEE J. Emerg. Sel. Top. Power Electron.* **2020**, *8*, 991–1003. [CrossRef]

35. Dimitrovski, R.; Mehlmann, G.; Luther, M.; Ndreko, M.; Winter, W.; Trunk, K.; Rachinger, S. Comprehensive computer program for SSTI analysis of AC/DC power systems based on the component connection method. *Int. J. Electr. Power Energy Syst.* **2019**, *111*, 508–516. [CrossRef]

36. Dragičević, T.; Lu, X.; Vasquez, J.C.; Guerrero, J.M. DC Microgrids—Part I: A Review of Control Strategies and Stabilization Techniques. *IEEE Trans. Power Electron.* **2016**, *31*, 4876–4891. [CrossRef]

37. Dragičević, T.; Lu, X.; Vasquez, J.C.; Guerrero, J.M. DC Microgrids—Part II: A Review of Power Architectures, Applications, and Standardization Issues. *IEEE Trans. Power Electron.* **2016**, *31*, 3528–3549. [CrossRef]

38. Mandrile, F.; Carpaneto, E.; Bojoi, R. VSG Simplified Damper Winding: Design Guidelines. In Proceedings of the IECON 2019—45th Annual Conference of the IEEE Industrial Electronics Society, Lisbon, Portugal, 14–17 October 2019; IEEE: Lisbon, Portugal, 2019; pp. 3962–3967. [CrossRef]

39. Teodorescu, R.; Liserre, M.; Rodriguez, P. *Grid Converters for Photovoltaic and Wind Power Systems*; John Wiley & Sons: Hoboken, NJ, USA, 2011.

40. Kjaer, P.C.; Gupta, M.; Martinez, A.; Saylors, S. Comparison of Experiences with Wind Power Plants with Low SCR. In Proceedings of the IEEE PES General Meeting, Denver, CO, USA, 26 March 2015; Technical Panel.

Article

Computationally Efficient Modeling of DC-DC Converters for PV Applications

Fabio Corti [1], Antonino Laudani [2], Gabriele Maria Lozito [2,*] and Alberto Reatti [1]

[1] Department of Information Engineering, University of Florence, Via di S.Marta 3, 50139 Florence, Italy;
fabio.corti@unifi.it (F.C.); alberto.reatti@unifi.it (A.R.)

[2] Engineering Department, Roma Tre University, Via Vito Volterra 62b, 00146 Roma, Italy;
alaudani@uniroma3.it

* Correspondence: gabrielemaria.lozito@uniroma3.it

Received: 26 August 2020; Accepted: 28 September 2020; Published: 30 September 2020

Abstract: In this work, a computationally efficient approach for the simulation of a DC-DC converter connected to a photovoltaic device is proposed. The methodology is based on a combination of a highly efficient formulation of the one-diode model for photovoltaic (PV) devices and a state-space formulation of the converter as well as an accurate steady-state detection methodology. The approach was experimentally validated to assess its accuracy. The model is accurate both in its dynamic response (tested in full linearity and with a simulated PV device as the input) and in its steady-state response (tested with an outdoor experimental measurement setup). The model detects automatically the reaching of a steady state, thus resulting in lowered computational costs. The approach is presented as a mathematical model that can be efficiently included in a large simulation system or statistical analysis.

Keywords: DC-DC converters; photovoltaics; single-diode model; state-space

1. Introduction

Modeling the dynamic behavior of a photovoltaic power system is a challenging and actual topic due to its nonlinear nature and its dependence on environmental quantities [1]. The whole conversion system is, in general, composed of several photovoltaic devices, DC-DC conversion stages, energy storage units, and, usually, an inverter for grid connection. Considering the chain from the end, grid-connected inverters are usually well-understood, and considerable literature can be found on modeling both their AC and transient responses [2–4]. Most researches consider parasitic components that highly affect the converter operation, using both a linear equivalent circuit approach and state-space averaging methods [5–13]. On the other hand, energy storage is an open topic considering the wide options for storage technologies that are available or rising [14,15]. The DC stage is considerably simple to model when the input voltage can be considered an ideal source; however, a nonlinear device such as a photovoltaic (PV) panel can void several assumptions used for the simple steady-state modeling of the DC converter. Electrical modeling of a PV device involves an interaction with the environmental variables of irradiance (G) and temperature (T) to correctly represent the voltage-current relationship of the device. The assessment and forecasting of such quantities, especially irradiance, is an open problem in the literature [16–19]. Modeling the PV device connected to a DC converter can be difficult, computationally intensive, and might lead to numerical instability. A complete overview of the issues related can be found in [1,4,20], where the stability of the control process is discussed in an exhaustive way, including the problematics related to maximum power point tracking and small-signal modeling of the PV source. The reading of [4,20] is also important to better understand the differences raised when the PV source is considered equivalent to a voltage or current source. The purpose of this work is to address the difficulties in modeling the nonlinearity of the PV source characteristic by discussing a computationally efficient implementation of the model. Two key aspects are optimized. The first one involves the simulation of the PV device by means of the single-diode model (in the version of a five-parameters model) [21,22]. This circuit model is accurate on the majority of silicon devices, under variable environmental conditions [23]. This individual element of the framework was chosen with solid foundations in state-of-the-art approaches. The single-diode model is a preferred choice for the vast majority of PV literature, both for its computational speed and the wide options for model identification algorithms available. The model was implemented in the framework, with some recent optimizations to enhance the numerical stability. Through the application of the Lambert W function, two explicit relationships can be formulated [24]: the voltage can be written as a function of the current or the current can be written as a function of the voltage, i.e., the PV source becomes equivalent to a current-controlled voltage source (CCVS) or to a voltage-controlled current source (VCCS). However, for some operating conditions of the one-diode model, an exponential overflow can cause divergence in the Lambert function expressions. This shortcoming is addressed through an additional functional mapping [25,26]. The second optimization involves the transient simulation of the DC-DC converter connected to the PV device. A procedure to predict the steady-state value of the state variables is proposed. Using this knowledge, the transient simulation can be halted as soon as a steady state is reached. It should be underlined that this work aims to create an efficient framework for the simulation of a DC-DC converter in the presence of a PV input; thus, the steady-state detection approach is its core concept, since it is able to shorten drastically the computational times. This is particularly important since a time domain analysis is performed to remark the nonlinear and time-varying behaviors of the PV source.

The accuracy of the proposed modeling approach was validated experimentally on three workbenches. The first one to assess the dynamic model of the DC-DC converter without the nonlinearity introduced by the PV device. The second to assess the accuracy of the steady-state predicting procedure when a real PV device is connected to the DC-DC converter. The third one to assess the accuracy of the transient response for the full DC-DC converter plus PV device system, making use of a PV device electrical emulator and a programmable DC load.

An important contribution comes from the presentation of the proposed approach. The majority of the simulations involving DC-DC converters and PV devices found in the literature make use of circuit simulation software [27–29]. The further inclusion of these simulations in larger analyses (e.g., Monte Carlo or a grid simulation for machine-learning purposes [30–32]) is difficult and unpractical. The approach proposed in this work, on the other hand, is presented in the form of clear mathematical models that can be implemented in any programming language.

The paper is structured as follows. In Section 2, the proposed model is described as composed by the PV device model, the transient and steady-state model for the DC-DC converter, and the steady-state prediction procedure, along with a computational costs discussion. In Section 3, the sizing and hardware implementation of the DC-DC converter is described. The three experimental validation workbenches are described, and the results are presented in Section 4. Final considerations and conclusions close the manuscript.

2. The Proposed Model

The model presented in this work represents a photovoltaic device connected to a DC-DC converter with a resistive load. The nature of the real system to represent includes both the electrical behavior of the DC-DC converter and the nonlinear, environmental-dependent nature of the PV device. The model aims to represent the electrical behavior of the full system with a good accuracy, with reduced computational costs both in the transient and steady state operations. This is achieved through a lean reformulation of the current-voltage relationship for a single-diode model, a state-space formulation for the DC-DC converter, and, most importantly, an effective strategy for steady-state detection.

2.1. Single-Diode Model for Transient Simulations

The single-diode model is an equivalent circuit model that can be used to represent the electrical behavior of a PV device with arbitrary conditions of irradiance (G) and temperature (T). The circuital representation of the model is shown in Figure 1 It should be noted that the single-diode model is meant to represent a single PV cell. However, under the assumptions of uniform irradiance and temperature, it can be used to represent an arbitrary series and parallel of PV cells. Thus, it is suitable to represent PV panels, strings, and arrays.

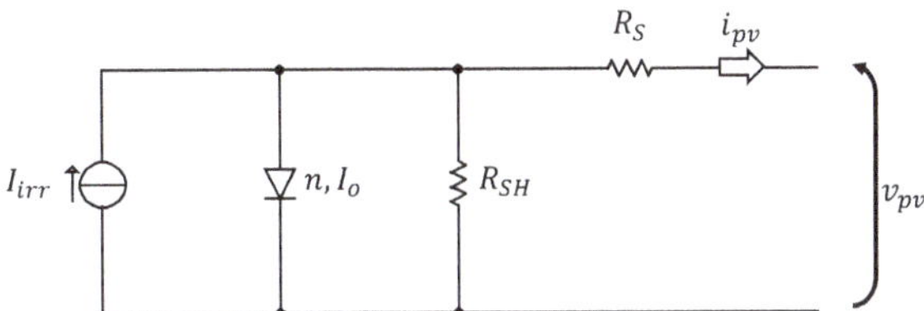

Figure 1. Single-diode equivalent circuit model for a photovoltaic (PV) device.

Considering the only part of the circuit where the voltage (v_{pv}) and current (i_{pv}) are defined in an active sign convention, the voltage-current relationship of this model can be easily derived by applying the Kirchhoff Current Law. As can be seen, it is nonlinear, transcendent, and implicit.

$$I_{irr} - I_o\left[\exp\left(\frac{v_{pv} + R_S\, I_{pv}}{n\; V_t}\right) - 1\right] - \frac{v_{pv} + R_S\, i_{pv}}{R_{SH}} - i_{pv} = 0 \tag{1}$$

The five parameters of the model are plainly apparent in the equation, where I_{irr} is the generated photocurrent, I_o is the diode reverse saturation current, n is the ideality factor, R_S is the series resistance, and R_{SH} is the shunt resistance. Since (1) describes the voltage-current relationship of the device, the general application involves the determination of the current for a known device voltage or vice-versa. Unfortunately, since (1) is implicit and transcendent, the only way to solve it is by using iterative root-finding techniques, such as the Newton-Raphson method, fixing either the voltage or the current, and solving the equation for the missing variable. This approach is very common and is acceptably fast but, as will be shown below, lacks a proper explicit formulation.

Regardless of the difficulty in solving (1), the single-diode model must be identified prior to being able to represent a physical device. Identification of the model (over a physical device) consists in the determination of the five parameters that results in the best fit with the electrical characteristic of the device at the standard reference conditions, SRC ($T = 25\ ^{\circ}$C and $G = 1000\ \text{W/m}^2$). Several methodologies exist in the literature that allow identification either from experimental measurements [25,33] or quantities from the constructor datasheet [22].

Since the circuit model of the device is identified at the SRC, the model must be completed by a set of equations [23] that express a dependence of the five circuit parameters from the environmental conditions of irradiance (G) and temperature (T).

$$R_S = R_{S,ref} \tag{2}$$

$$n = n_{ref} \tag{3}$$

$$R_{sh} = R_{sh,ref}\left(\frac{G}{G_{ref}}\right) \tag{4}$$

$$I_{irr} = I_{irr,ref}\left(\frac{G_{ref}}{G}\right)\left[1 + \alpha\left(T - T_{ref}\right)\right] \tag{5}$$

$$I_0 = I_{0,ref}\left(\frac{T}{T_{ref}}\right)^3 \exp\left[\frac{E_g(T_{ref})}{kT_{ref}} - \frac{E_g(T)}{kT}\right] \tag{6}$$

Through this set of equations, it is possible to update the electrical characteristic of the model for a time-variable profile of irradiance and temperature, which is the classic scenario for a real PV device. The subscript *ref* is for the parameters when the model represents the device at the SRC. The other symbols are k for the Boltzmann constant, α for the temperature coefficient for the open-circuit current, and $E_g(T)$ is the temperature-dependent bandgap energy of the semiconductor used for the device construction. If this current-voltage (I-V) relationship is to be included in a transient analysis, such as the one for a DC-DC converter, for each time step, a nonlinear root-finding problem should be solved, with an added computational burden. This is especially true if several PV devices are connected in a series or in parallel. A first reformulation of this relationship is found in the literature through the means of the Lambert W function, resulting in two expressions [24]:

$$v_{pv}(i_{pv}) = R_{SH}(I_{irr} + I_0) - (R_S + R_{SH})i_{pv} - nV_t\; W\left[\frac{I_0 R_{SH}}{nV_t}\exp\frac{R_{SH}(I_{irr} + I_0 - i_{pv})}{nV_t}\right] \tag{7}$$

$$i_{pv}(v_{pv}) = -\frac{nVt}{R_S}W\left[\frac{R_S}{nV_t} * \frac{I_0 R_{SH}}{R_{SH} + R_S} * \exp\left(\frac{R_{SH}}{R_S + R_{SH}}\frac{v_{pv} + R_S(I_{irr} + I_0)}{nV_t}\right)\right] + \frac{(I_{irr} + I_0)R_{SH} - v_{pv}}{R_{SH} + R_S} \tag{8}$$

This reformulation moves the computational burden of the root-finding algorithm from the solution of a generic equation to the solution of the Lambert W function for a positive argument. Although the explicit current relationship (8) is, in general, without computational problems, the same cannot be said for the explicit voltage relationship [25,26]. When very large arguments appear in the exponential used in (7), which is limited to arguments below ~700 in a 64-bit double-precision float computation environment, the overflow may lead to divergence. A possible solution involves using a mapping function such as the one in (9) and (10).

$$g(z) = \ln[W\cdot\exp(z)] \tag{9}$$

$$g(z) + \exp[g(z)] = z \tag{10}$$

From this mapping, the explicit voltage relationship is

$$v_{pv}(i_{pv}) = -i_{pv}\cdot R_S + n\cdot V_t\left[g(z) - \ln\left(\frac{I_0 R_{SH}}{nV_t}\right)\right] z = \ln\left(\frac{I_0 R_{SH}}{nV_t}\right) + \frac{R_{SH}(I_{irr} + I_0 - i_{pv})}{nV_t} \tag{11}$$

which can be used along with (8) to include the I-V relationship of a PV device in any Kirchhoff circuital equation resulting from a state-space analysis of a dynamic circuit.

2.2. Transient Model for the DC-DC Converter

The topology investigated in this work is a classic buck-boost converter. This topology is useful both for load-oriented applications (e.g., battery charging and equalization [34,35]) and source-oriented applications (e.g., maximum power point tracking [36]). The circuit model, featuring the main parasitic elements, is shown in Figure 2 where, C_{IN} is a capacitor stabilizing the converter input voltage, R_{ds} is the on resistance of the controlled switch (e.g. a power MOSFET), R_L is the inductor L equivalent series resistance, V_{fwd} is the diode threshold voltage, R_D is the diode conduction resistance, C is the converter output capacitance and R is the converter load resistance; moreover, i_{pv} is the panel output current, i_L is the inductor current, v_{CIN} is the voltage across the input capacitor and, v_C is the voltage across the output. Since the purpose of this work is to create a model able to represent the full system both in the transient and steady states, the first step involves the identification of the state variables of the system and the derivation of the relative state equations. The system is commuted through a switch and a diode, and for this reason, two sets of state equations needs to be defined: one for the ON state and one for the OFF state. Through simple observations, the ON and OFF state equations of the circuit can be derived.

$$
ON: \begin{cases} \dot{x}_1 = \frac{1}{C}\left(-\frac{x_1}{R}\right) \\ \dot{x}_2 = \frac{1}{L}[x_3 - (R_{DS} + R_L)x_2] \\ \dot{x}_3 = \frac{1}{C_{IN}}\left[i_{pv}(x_3) - x_2\right] \end{cases} \tag{12}
$$

$$
OFF: \begin{cases} \dot{x}_1 = \frac{1}{C}\left(-\frac{x_1}{R} - x_2\right) \\ \dot{x}_2 = \frac{1}{L}\left(-x_2 R_L + x_1 - V_{fwd} - R_D x_2\right) \\ \dot{x}_3 = \frac{1}{C_{IN}}\left[i_{pv}(x_3)\right] \end{cases} \tag{13}
$$

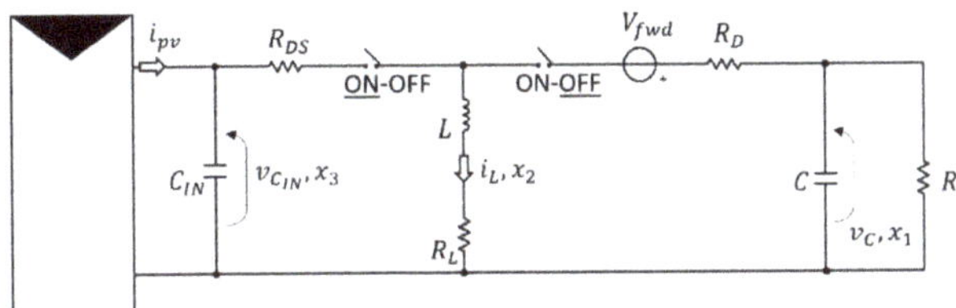

Figure 2. Circuit model for the PV device connected to the DC-DC converter, highlighting the parasitic components.

The state variables appearing in (12) and (13) are the following: x_1 is the voltage across the output capacitor, x_2 is the current through the inductor, and x_3 is the voltage across the input capacitor. As can be seen in the last state equation, the I-V relationship of the PV device affects the current of the input capacitor. This is the only nonlinearity present in the dynamic model of the system. The PV device is connected directly in parallel with the input capacitor, and since its voltage is known at each time step (being a state variable), the most convenient choice to express the PV electrical relationship is (8), and consequently, it acts as the VCCS. However, if a more complex converter was implemented, or if, for some reason, the input capacitor was to be removed, the series connection between the PV device and the main inductor L would result in a more practical choice of (11) as the relationship to represent the PV device.

The state equations can be integrated numerically to create a time response of the system. Assuming a variable profile for irradiance and temperature, the PV circuit parameters (and thus, the resulting state equations) can be updated by means of (2)–(6). As the system reaches a steady state, the integration can halt, and the simulation can proceed with a steady-state linearized model. Since the system is nonlinear, it is impossible to estimate the steady state from the time constants, and it must be manually detected by observing the state variables.

2.3. Steady-State Detection and Modeling

In a steady state, it is safe to assume that the large capacitor in parallel with the PV device has a stable and constant voltage across it. From a system point of view, this transforms the former state variable in a system input. Since the nonlinearity is hidden behind this input, it is possible to approach the DC-DC converter simulation with classic tools such as defining the transfer function via the state-space average (SSA) [37,38]. With reference to the first two equations of (12) and (13), the linearized and averaged system is shown in (14), where V_{IN} is the constant voltage across the C_{IN} input capacitor. This new system can be integrated analytically or numerically with a large time step for reduced computational costs.

$$
\overline{A} = \begin{bmatrix} -\frac{1}{RC} & \frac{D-1}{C} \\ \frac{1-D}{L} & \frac{D(R_D - R_{DS}) - (R_L + R_D)}{L} \end{bmatrix}; \quad \overline{B} = \begin{bmatrix} 0 & 0 \\ \frac{D}{L} & \frac{D-1}{L} \end{bmatrix}; \quad u = \begin{bmatrix} V_{IN} \\ V_{fwd} \end{bmatrix}; \quad x = \begin{bmatrix} x_1 \\ x_2 \end{bmatrix} \dot{x} = \overline{A}x + \overline{B}u \tag{14}
$$

From the SSA system, by assuming a steady state, three transfer functions can be easily derived:

$$\frac{V_{OUT}}{V_{IN}} = \frac{-D}{(D-1) + \frac{D(R_D - R_{DS}) - (R_L + R_D)}{R(1-D)}} \tag{15}$$

$$\frac{I_{OUT}}{I_{IN}} = \frac{1-D}{D} \tag{16}$$

$$\frac{R_{OUT}}{R_{IN}} = \frac{-D^2}{(1-D)\left[(D-1) + \frac{D(R_D - R_{DS}) - (R_L + R_D)}{R(1-D)}\right]} \tag{17}$$

Indeed, in (17), the R on the right-hand side of the equation is equal to R_{OUT}. The transfer functions can be used as a method to detect the steady state through these simple steps:

- Using (17), find the equivalent input load seen by the PV panel across the DC-DC converter.
- Find the operating point of the PV panel through (11) or (8) and the steady-state input voltage $V_{IN,SS}$.
- Using (15), determine the steady-state output voltage $V_{OUT,SS}$.
- Calculate the output current with Ohm's law, then use (16) to determine the input current $I_{IN,SS}$.
- Calculate the average current through the inductor as $I_{L,SS} = I_{IN,SS}/D$.
- To determine the accuracy of the steady state, compare x_1 with $V_{OUT,SS}$, x_2 with $I_{IN,SS}$, and x_3 with $V_{IN,SS}$.

An example of this comparison is shown in Figure 3 for three different values of duty cycle D (0.4, 0.5, and 0.6). The transient reaches a steady-state value that is very close to the predicted steady-state value obtained through the (15)–(17) transfer functions. From an implementation point of view, since the electric quantities are averagely constant once a steady state is reached, the simulation can be halted or, if needed, can be performed with very large time steps using (14), which can hold for small perturbations as well. This results in an almost negligible computational cost. On the other hand, if a large perturbation (e.g., on the load or in the environmental conditions of the PV device) occurs, it is possible to restart the transient simulation using the steady-state quantities of $V_{OUT,SS}$, $I_{L,SS}$, and $V_{IN,SS}$ as the initial conditions for the three state variables: x_1, x_2, and x_3.

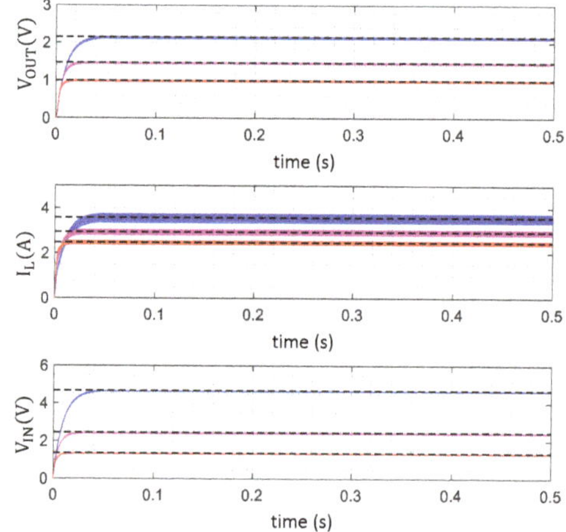

Figure 3. Comparison between the transient solution of the circuit (full-colored lines) and the estimated steady-state value of the three state variables (dashed black lines). The blue curve is relative to $D = 0.4$, magenta to $D = 0.5$, and red to $D = 0.6$.

2.4. Computational Costs

The model computational load is only relevant for the transient part and is composed by the short time-step integration of (12) and (13). Inside this computation, a large contribution to the computational cost comes from the solution of (8) by means of the Lambert W function at each time step. The model was profiled for performance on a Core i7 Windows 10 machine running Matlab2020a by The Mathworks, Inc., Natick, Massachusetts (MA)considering a variable number of time steps (from 6400 to 640,000) to account for the memory usage effects. The average computational time for a single time step is 4.213×10^{-5} s. The contribution from the solution of (8), considering the variable irradiance and temperature (thus, considering the use of (4)–(6) for parameter updates) is 1.731×10^{-5} s. In other words, more than 40% of the computational burden is related to the nonlinear nature of the problem and, in particular, to the solution of the Lambert W function found in (8) (or the g function found in (11), in case a voltage relation is needed).

The costs involved for the transient part of the model are considerable and would lead to long computational times for simulations. In this context, the ability to quickly switch into a simpler steady-state model after the transient is exhausted is a strong feature that makes this approach suitable for further integrations into more complex analysis protocols, such as Monte Carlo optimizations.

3. DC-DC Converter

To validate the proposed model, a DC-DC converter was built to interface the load with the photovoltaic panel. The converter constraints are summarized in Table 1.

Table 1. System parameters.

Parameter	Value	Description
f_0	20 kHz	Switching Frequency
D	0.3–0.6	Duty Cycle Range
$V_{i,max}$	33.3 V	Maximum Input Voltage
I_o	6.67–40 A	Output Current
R_L	0-20 Ω	Load Resistance
ΔV_o	0.3 V	Maximum Output Ripple
P_o	100–295 W	Output Power

The procedure given in [8,39] is used for the DC-DC converter design. The average output current is calculated as

$$I_o = \frac{I_o^{\max} + I_o^{\min}}{2} = 23.34 \text{ A} \tag{18}$$

The inductor average current is expressed as

$$I_L = \frac{I_o}{1 - D} \tag{19}$$

Thus, the inductor average current I_L depends on the duty cycle. If the system operates at $D = D_{\max} = 0.6$, the current is $I_L^{\max} = 58.34$ A, while if the converter operates at $D = D_{\max} = 0.3$, the current is $I_L^{\min} = 33.34$ A.

The minimum value of inductance needed to operate in continuous conduction mode (CCM) is

$$L_{\min} = \frac{R_L^{\max} (1 - D_{\min})^2}{2f} = 184 \text{ }\mu\text{H} \tag{20}$$

while the minimum capacitance needed to operate at the desired output voltage ripple is

$$C_{min} = \frac{I_0^{max} D_{min}}{f\left[\Delta V_o - \left(\frac{I_0^{max}}{1-D} + \frac{\Delta I_L}{2}\right)ESR_C\right]} = 578\ \mu F \tag{21}$$

The input capacitance is used to hold up the input voltage during the time when the energy is decreasing in the inductor.

If the input voltage drop should not be bigger than the input voltage ripple ΔV_{in}, the minimum effective value for this capacitor C_{IN}^{min} is estimated with

$$C_{IN}^{min} = \frac{I_L D^{max}}{f\left[\Delta V_{in} - \Delta I_L ESR_{C_{IN}}\right]} = 1.73\ mF \tag{22}$$

Equation (10) in [40] implies that higher equivalent series resistance of the capacitor (ESR) increases the input voltage drop. A SCT3022AL SiC power MOSFET driven by a 1EDC60H12AH isolated gate driver is used in the Buck-Boost converter.

The MOSFET has a breakdown voltage $V_{DSS} = 650$ V and a nominal conduction resistance of $r_{DS} = 22$ mΩ. A FFSP2065B SiC diode with a reverse voltage $V_{RRM} = 650$ V and a continuous rectified forward current $I_F = 20$ A is used. The final sizing for the dynamic components of the DC-DC converter used in the experimental validations is shown in Table 2. It should be noted that the ESR_{CIN} and ESR_C were reported in the table, but for simplicity, they were neglected in the circuit model.

Table 2. Dynamic component values.

Component	Measured
Inductance L	224.62 μH/$ESR_L = 0.023\ \Omega$
Capacitance C	662.32 μF/$ESR_C = 0.016\ \Omega$
Input Capacitance C_{IN}	2937.2μF/$ESR_{CIN} = 0.016\ \Omega$

4. Experimental Validation

The converter described in Section 3 was utilized in three different experimental workbenches (WB): WB1 involved the measurement of the dynamic response for the DC-DC converter at a constant voltage input. WB2 involved the steady-state response of the DC-DC converter with a real PV panel as an input and operated under outdoor variable environmental conditions. WB3 involved the dynamic response for the DC-DC converter with a hardware-simulated PV device as the input and a programmable load as an output. The experimental measurements of each WB are compared against the simulation based on the model described in Section 2.

4.1. WB1: Constant Voltage Input

The purpose of the first test is to assess the transient accuracy of the model for the DC-DC converter by itself, using as input a voltage source. As stated before, a PV source is drastically different from a voltage source, but we consider this first test important for the setup of the approach. By removing the main nonlinearity from the system (i.e., the PV device), this allows for an analysis that is focused only on the transient evolution of the fast-dynamic elements of the converter. The measurements and the simulations represent the turn-on transient of the DC-DC converter with a constant input voltage. The transient measurements were acquired using a dSPACE MicroLabBox (dSPACE Inc. Wixom, MI, USA), which is an integrated solution for measurement and hardware development. The system is based on an NXP (NXP Semiconductors Netherlands, Eindhoven) processor (dual core, 2 GHz), features eight analog input channels with independent ADC conversion (14-bit and 10 Msps) and a voltage range of ±10 V. The system features analog outputs as well, with 16 channels featuring independent DAC conversion (16-bit and 1 Msps). Moreover, the system features 48 digital I/O ports configurable to implement the most common digital interfaces (e.g., CAN, I2C, SPI, etc.). For the purpose of this setup, the digital output of the dSPACE was used to create the driving signal for the Pulse Width Modulation (PWM) of the DC-DC converter, and the analog inputs were used to sample the input and output voltage of the converter. The results of the output voltage at constant input voltages of 5 V and 7 V are shown in Figure 4.

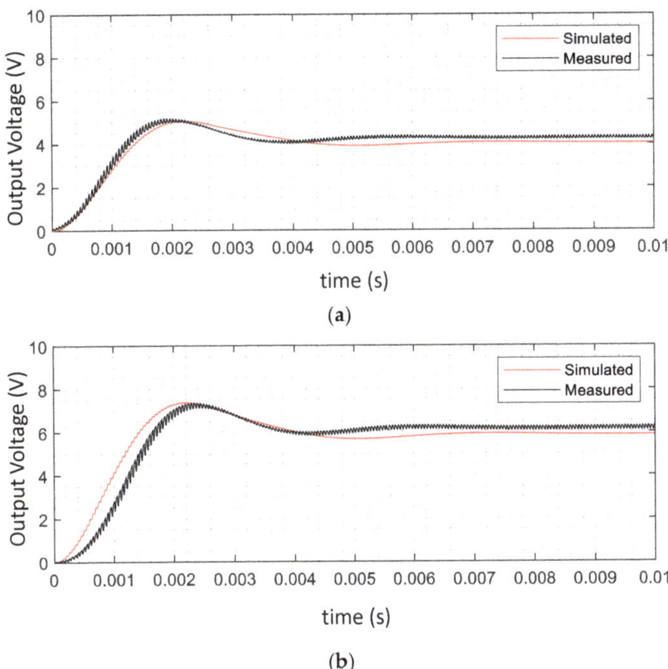

Figure 4. Turn-on transient response of the converter with constant input voltages: 5 V (**a**) and 7 V (**b**).

4.2. WB2: Outdoor PV

The purpose of the second test is to assess the steady-state accuracy of the model when a real PV panel is utilized as a source. The steady-state value is strongly influenced by the environmental values of irradiance and temperature. For this reason, an outdoor setup was implemented featuring a 230-W PV Panel TW230P60-FA2 by Tianwei, Changhua, Taiwan(a datasheet extract showing the open circuit voltage V_{OC}, the short circuit current I_{SC}, the maximum power voltage V_{MP} and the maximum power current I_{MP} can be seen in Table 3) and an instrumental chain for the acquisition of instantaneous irradiance, panel backside temperature, DC-DC input voltage, and DC-DC output voltage. The measurements are performed using a dedicated instrument for PV panel testing, the Chauvin Arnoux Green Test FTV100 by Chauvin Arnoux, Paris, France. This instrument features input ports for both AC and DC voltage and the current measurements, along with a set of input ports for the temperature and irradiance probes. For the purpose of this test, the DC ports for voltage (up to 1000 V) and current (up to 200 A) were used. The temperature was measured through a PT100 temperature probe by Chauvin Arnoux, Paris, France (−30 to 80 °C range with 1% accuracy), and the solar irradiance was measured through a pyranometer by Chauvin Arnoux, Paris, France (up to a 2000 W/m^2 range with 2% accuracy). For this measurement, a constant load resistance 11 Ω and three different duty cycles (40%, 50%, and 60%) were used; the choice was done to test the system in a mid-working condition. Measurements were performed in January 2020 at 43.799° N. The measurement chain is schematized in Figure 5, along with the electrical characteristics of the panel in Figure 6.

The field implementation is shown in Figure 7. The irradiance values and temperature (Figure 8) recorded by the FTV100 were used to simulate the PV device under variable conditions. The Chauvin Arnoux was configured to average the temperature and irradiance measurements over 10 s. Thus, 300 points is a measurement spanning over 50 min. The small irradiance values are a consequence of the local weather during the tests. It is worth noticing that obtaining a good steady-state accuracy at low irradiances is much more difficult than at quasi-SRC, due to the behavior of the R_{SH} parameter of the one-diode model for low values of G. The steady-state input and output voltages, simulated and measured, are shown in Figure 9.

Table 3. TW230P60 electrical characteristics.

V_{OC} (V)	I_{SC} (A)	V_{MP} (V)	I_{MP} (A)
37.3	8.22	29.4	7.82

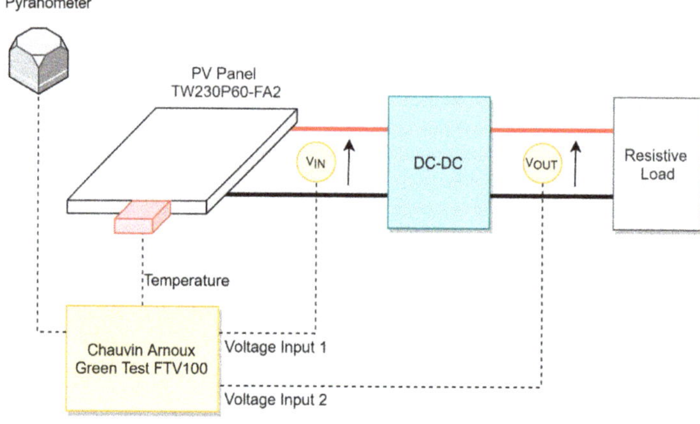

Figure 5. Measurement chain for the steady-state response over a variable irradiance and temperature.

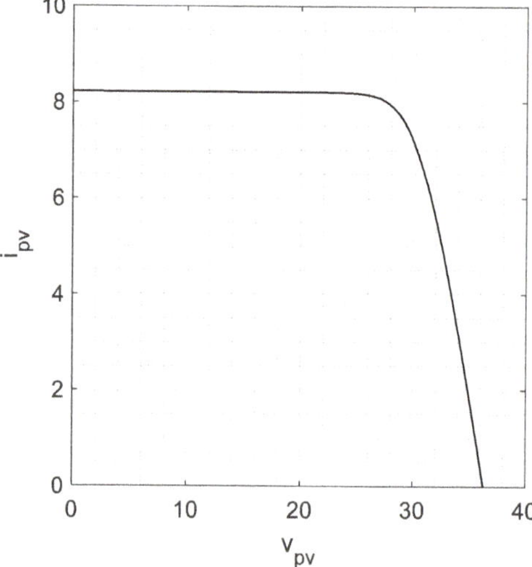

Figure 6. Electrical characteristics of the TW230P60-FA2 panel at the standard reference conditions (SRC).

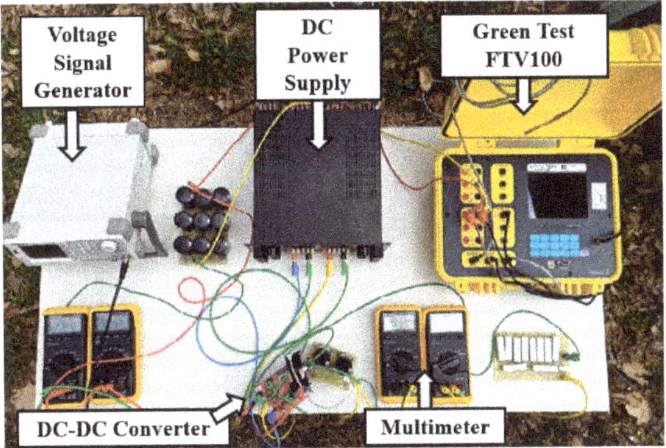

Figure 7. Experimental workbench for the outdoor PV test.

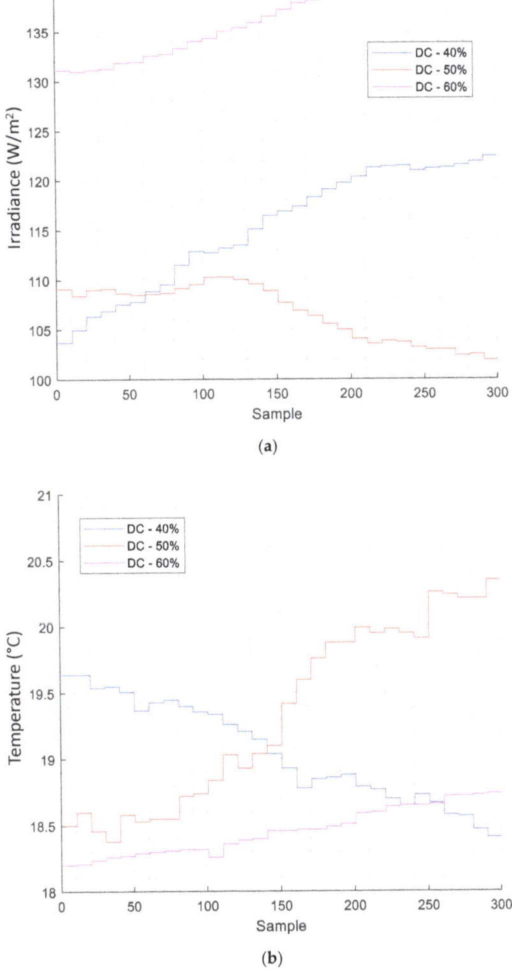

(a)

(b)

Figure 8. Irradiance (**a**) and temperature (**b**) conditions recorded during the measurement.

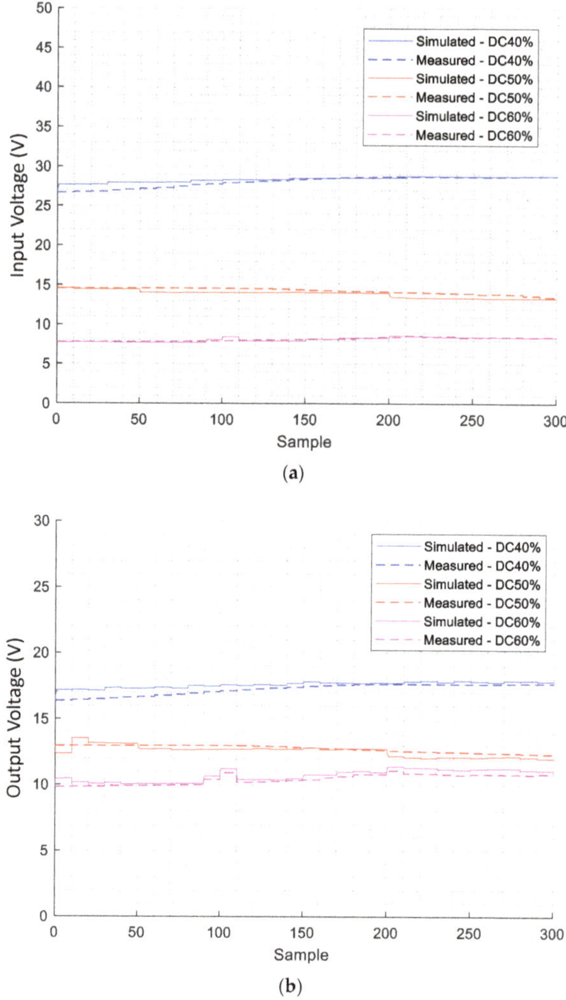

Figure 9. Simulated and measured steady-state input (**a**) and output (**b**) voltages.

4.3. WB3: Variable Output Load

The purpose of the third test is to assess the transient accuracy of the model when a PV source is present. Since changes in the irradiance and temperature are, in general, very slow if compared to the time constants of the system, there is no real interest in simulating the transient involved in them. On the other hand, the output load can change abruptly if a sudden energy sink is present. For the purpose of the experimental validation, this scenario was created by using a step-variable load. The measurement chain, schematized in Figure 10 and shown in Figure 11, is composed by a four-channel oscilloscope (Tektronix TDS3014b by Tektronix, Inc. Beaverton, OR, USA), a 700-W PV Device programmable simulator (TerraSAS ETS60), and a programmable AC-DC load (Itech IT 8615). A summary of the instrument's relevant characteristics is shown in Table 4. In particular, the PV simulator was programmed to simulate the panel reported in Table 5. The programmable load was programmed to switch between the values of 7.5 Ω and 17.5 Ω. The choice for these values was made considering that the PV device would work at the SRC, and for this I-V relationship, the panel would switch from a quasi-SC (Short Circuit) condition (at 7.5 Ω) to a quasi-OC (Open Circuit) condition (at 17.5 Ω), achieving a large transient. The duty cycle of the DC-DC converter was kept at $D = 0.5$. The results of the two tests are shown in Figures 12 and 13. In the first one, the load resistance initially is 17.5 Ω and is commuted to 7.5 Ω once a steady state is reached. In the second one, the initial load resistance is 7.5 Ω and is commuted to 17.5 Ω before the steady state is reached. All the performed experimental tests confirm that the model correctly simulates both the transient and steady-state operations of the considered systems.

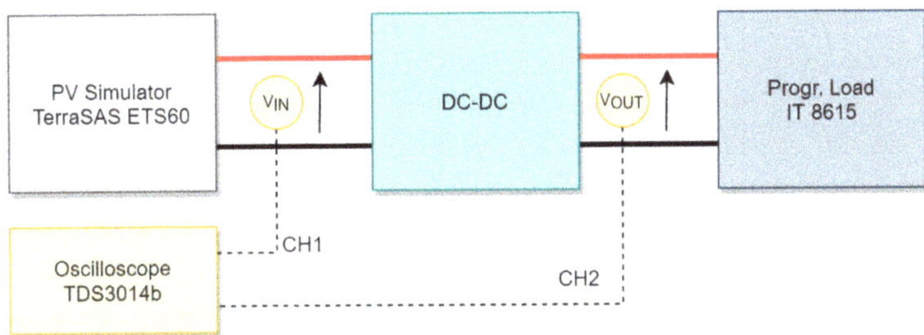

Figure 10. Measurement chain for the transient response over a step-variable load change.

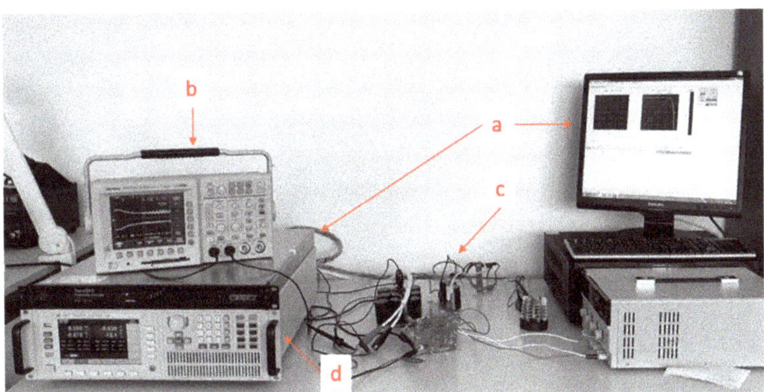

Figure 11. Experimental workbench depicting the TerraSAS ETS60 (**a**), the TDS3014b (**b**), the DC-DC converter (**c**), and the IT8615 (**d**).

Table 4. Instrumental chain of the workbench 3 (WB3) summary.

Instrument	Characteristics
Tektronix TDS3014b Oscilloscope	Four channels, up to 100 MHz 1 mV/div to 10 V/div vert. sensitivity 4 ns/div minimal time base GPIB and LAN connectivity
Itech IT 8615 Electronic Load	Up to 420 Vrms range and 1800 VA 45-Hz to 450-Hz frequency range GPIB, LAN, and USB connectivity
TerraSAS ETS60 Solar Simulator	Output up to 66 V (OC) and 14 A (SC) Maximum output power 714 W I-V curve resolution of 1024 points LAN connectivity

Table 5. Parameters for the PV simulator.

V$_{OC}$ (V)	I$_{SC}$ (A)	V$_{MP}$ (V)	I$_{MP}$ (A)
20	1.10	17.22	1.04

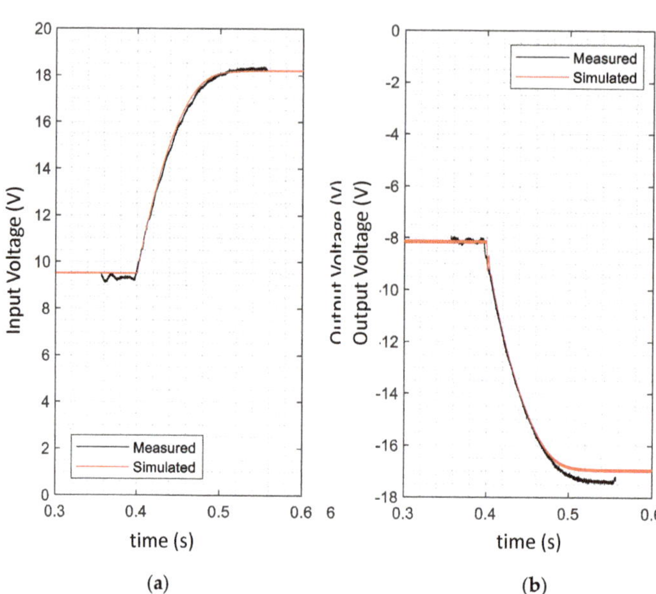

Figure 12. Transient evolution of the input voltage (**a**) and output voltage (**b**) for an output load change from 7.5 Ω to 17.5 Ω.

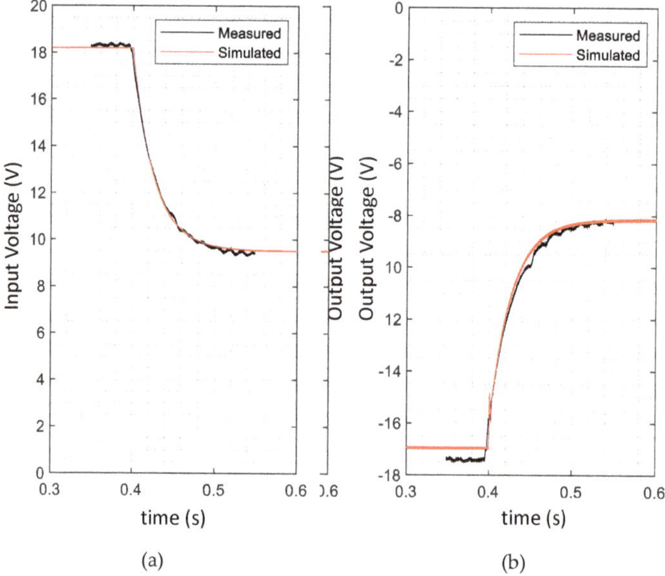

Figure 13. Transient evolution of the input voltage (**a**) and output voltage (**b**) for an output load change from 17.5 Ω to 7.5 Ω.

The results obtained from this last workbench confirm the accuracy of the dynamic model even in presence of the PV device nonlinearity. Although a small error can be seen in the steady-state parts (which can be accounted for considering the nonideal connections), the transient evolution timings match almost perfectly the simulation. This is particularly meaningful if this model is to be used for the study of the DC-DC converter control systems, such as an MPPT controller, where an incorrect estimate of the time required for the system response could result in a completely different behavior between the simulations and the real system.

5. Conclusions

In this paper, an efficient and accurate methodology to simulate a DC-DC converter in photovoltaic applications was proposed, with the aim of creating a framework suitable for further integration in larger simulation and analysis environments. The proposed model was formulated to represent both the physical-to-electrical behavior of a photovoltaic device (i.e., taking into account nonelectrical quantities such as temperature and irradiance) and the dynamic nature of the DC-DC converter. The accuracy of the proposed approach was validated against three different experimental workbenches (two for the transient response and one for the steady-state response) to assess the reliability of the approach. The model in the steady state can be integrated with state-of-the-art measurements and the forecasting of irradiance [17–19] for power plant-produced energy estimations, and, in general, to assess the electrical quantities of the system over a large timespan. The dynamic model is crucial for innovative photovoltaic applications, such as automotive [41], where sudden changes in the environmental quantities often perturb the steady state, or PV systems, with highly variable power absorptions.

The possibility of integration for such a model in larger structures, which is the main aim of this research, is related to its computational efficiency and the wide adaptability that comes from its constituent parts. The single-diode model might not be the most accurate model for PV devices in the literature (more accurate, yet slower, is the double-diode model [42]), but it is a standard de facto for the representation of silicon devices, has solid literature concerning the model identification, and is computationally light. Moreover, the state-space approach, and the explicit current-voltage characteristics introduced for the PV device, make the methodology described in this work applicable to many different DC-DC converter topologies.

The methodology has, indeed, possible evolution and open problems. The choice of the single-diode model was widely justified for silicon devices, yet for some new technologies such as organic PV, it would be interesting to implement and validate other, more complex, circuit models [43,44]. To make this inclusion viable from a computational point of view, it might require numerical optimization of the circuit relations describing the models [45,46].

A second interesting possibility is to further extend the model chain. On the PV side, the model could be coupled with forecasting or measurement methodologies for the environmental quantities of irradiance and temperature [17–19]. On the load side, grid connection could be considered along with a load profile. This later quantity, analogously to the environmental quantities of irradiance and temperature, has a large literature field concerning measurement and forecasting [47–49].

Author Contributions: Conceptualization and supervision, A.L. and A.R.; methodology, A.L. and G.M.L.; investigation, G.M.L. and F.C.; validation, G.M.L. and A.L.; and prototyping and measurements, A.R. and F.C., manuscript writing F.C. and G.M.L., manuscript revision A.L. and A.R. All authors have read and agreed to the published version of the manuscript.

Funding: This research received no external funding.

Conflicts of Interest: The authors declare no conflict of interest.

References

1. Nousiainen, L.; Puukko, J.; Mäki, A.; Messo, T.; Huusari, J.; Jokipii, J.; Viinamäki, J.; Lobera, D.T.; Valkealahti, S.; Suntio, T. Photovoltaic generator as an input source for power electronic converters. *IEEE Trans. Power Electron.* **2013**, *28*, 3028–3038. [CrossRef]
2. Agorreta, J.L.; Borrega, M.; López, J.; Marroyo, L. Modeling and control of N-paralleled grid-connected inverters with LCL filter coupled due to grid impedance in PV plants. *IEEE Trans. Power Electron.* **2011**, *26*, 770–785. [CrossRef]
3. Figueres, E.; Garcerá, G.; Sandia, J.; González-Espín, F.; Rubio, J.C. Sensitivity study of the dynamics of three-phase photovoltaic inverters with an LCL grid filter. *Proc. IEEE Trans. Ind. Electron.* **2009**, *56*, 706–717. [CrossRef]
4. Suntio, T.; Leppäaho, J.; Huusari, J.; Nousiainen, L. Issues on solar-generator interfacing with current-fed MPP-tracking converters. *IEEE Trans. Power Electron.* **2010**, *25*, 2409–2419. [CrossRef]
5. Luchetta, A.; Manetti, S.; Piccirilli, M.C.; Reatti, A.; Kazimierczuk, M.K. Comparison of DCM operated PWM DC-DC converter modelling methods including the effects of parasitic components on duty ratio constraint. In Proceedings of the 2015 IEEE 15th International Conference on Environment and Electrical Engineering, Rome, Italy, 10–13 June 2015.
6. Davoudi, A.; Jatskevich, J.; Chapman, P.L. Averaged modelling of switched-inductor cells considering conduction losses in discontinuous mode. *IET Electr. Power Appl.* **2007**, *1*, 402–406. [CrossRef]
7. Luchetta, A.; Manetti, S.; Piccirilli, M.C.; Reatti, A.; Kazimierczuk, M.K. Effects of parasitic components on diode duty cycle and small-signal model of PWM DC-DC buck converter in DCM. In Proceedings of the 2015 IEEE 15th International Conference on Environment and Electrical Engineering, Rome, Italy, 10–13 June 2015.
8. Kazimierczuk, M.K. *Pulse-Width Modulated DC-DC Power Converters*, 2nd ed.; John Wiley & Sons, Inc.: New York, NY, USA, 2015; ISBN 9781119009524.
9. Davoudi, A.; Jatskevich, J.; Chapman, P.L. Simple method of including conduction losses for average modelling of switched-inductor cells. *Electron. Lett.* **2006**, *42*, 1246–1248. [CrossRef]
10. Amir, S.; Van Der Zee, R.; Nauta, B. An improved modeling and analysis technique for peak current-mode control-based boost converters. *IEEE Trans. Power Electron.* **2015**, *30*, 5309–5317. [CrossRef]
11. Davoudi, A.; Jatskevich, J.; Chapman, P.L. Numerical dynamic characterization of peak current-mode-controlled DC-DC converters. *IEEE Trans. Circuits Syst. II Express Briefs* **2009**, *56*, 906–910. [CrossRef]
12. Reatti, A.; Balzani, M. Computer aided small-signal analysis for PWM DC-DC converters operated in discontinuous conduction mode. In Proceedings of the Midwest Symposium on Circuits and Systems, Covington, KY, USA, 7–10 August 2005.
13. Ayachit, A.; Reatti, A.; Kazimierczuk, M.K. Small-signal modeling of PWM dual-SEPIC DC-DC converter by circuit averaging technique. In Proceedings of the IECON Proceedings (Industrial Electronics Conference), Florence, Italy, 23–26 October 2016.
14. Díaz-González, F.; Sumper, A.; Gomis-Bellmunt, O.; Villafáfila-Robles, R. A review of energy storage technologies for wind power applications. *Renew. Sustain. Energy Rev.* **2012**, *16*, 2154–2171. [CrossRef]
15. González, A.; Goikolea, E.; Barrena, J.A.; Mysyk, R. Review on supercapacitors: Technologies and materials. *Renew. Sustain. Energy Rev.* **2016**, *58*, 1189–1206. [CrossRef]
16. Lorenz, E.; Hurka, J.; Heinemann, D.; Beyer, H.G. Irradiance Forecasting for the Power Prediction of Grid-Connected Photovoltaic Systems. *IEEE J. Sel. Top. Appl. Earth Obs. Remote Sens.* **2009**, *2*, 2–10. [CrossRef]
17. Diagne, M.; David, M.; Lauret, P.; Boland, J.; Schmutz, N. Review of solar irradiance forecasting methods and a proposition for small-scale insular grids. *Renew. Sustain. Energy Rev.* **2013**, *27*, 65–76. [CrossRef]
18. Dong, Z.; Yang, D.; Reindl, T.; Walsh, W.M. Short-term solar irradiance forecasting using exponential smoothing state space model. *Energy* **2013**, *55*, 1104–1113. [CrossRef]
19. Oliveri, A.; Cassottana, L.; Laudani, A.; Riganti Fulginei, F.; Lozito, G.M.; Salvini, A.; Storace, M. Two FPGA-Oriented High-Speed Irradiance Virtual Sensors for Photovoltaic Plants. *IEEE Trans. Ind. Inform.* **2017**, *13*, 157–165. [CrossRef]
20. Suntio, T.; Messo, T.; Aapro, A.; Kivimäki, J.; Kuperman, A. Review of PV generator as an input source for power electronic converters. *Energies* **2017**, *10*, 1076. [CrossRef]

21. Catelani, M.; Ciani, L.; Kazimierczuk, M.K.; Reatti, A. Matlab PV solar concentrator performance prediction based on triple junction solar cell model. *Meas. J. Int. Meas. Confed.* **2016**, *88*, 310–317. [CrossRef]

22. Laudani, A.; Riganti Fulginei, F.; Salvini, A. Identification of the one-diode model for photovoltaic modules from datasheet values. *Sol. Energy* **2014**, *108*, 432–446. [CrossRef]

23. De Soto, W.; Klein, S.A.; Beckman, W.A. Improvement and validation of a model for photovoltaic array performance. *Sol. Energy* **2006**, *80*, 78–88. [CrossRef]

24. Corless, R.M.; Gonnet, G.H.; Hare, D.E.G.; Jeffrey, D.J.; Knuth, D.E. On the Lambert W function. *Adv. Comput. Math.* **1996**, *5*, 329–359. [CrossRef]

25. Blakesley, J.C.; Castro, F.A.; Koutsourakis, G.; Laudani, A.; Lozito, G.M.; Riganti Fulginei, F. Towards non-destructive individual cell I-V characteristic curve extraction from photovoltaic module measurements. *Sol. Energy* **2020**, *202*, 342–357. [CrossRef]

26. Roberts, K.; Valluri, S.R. Solar Cells and the Lambert W Function. In *Celebrating 20 Years of the Lambert W Function*; London, ON, Canada, 2016.

27. Abdulkadir, M.; Samosir, A.S.; Yatim, A.H.M. Modelling and simulation of maximum power point tracking of photovoltaic system in Simulink model. In Proceedings of the 2012 IEEE International Conference on Power and Energy (PECon), Kota Kinabalu, Malaysia, 2–5 December 2012; pp. 325–330.

28. Suskis, P.; Galkin, I. Enhanced photovoltaic panel model for MATLAB-simulink environment considering solar cell junction capacitance. In Proceedings of the IECON 2013-39th Annual Conference of the IEEE Industrial Electronics Society, Vienna, Austria, 10–13 November 2013; pp. 1613–1618.

29. Kharb, R.K.; Shimi, S.L.; Chatterji, S.; Ansari, M.F. Modeling of solar PV module and maximum power point tracking using ANFIS. *Renew. Sustain. Energy Rev.* **2014**, *33*, 602–612. [CrossRef]

30. Takruri, M.; Farhat, M.; Barambones, O.; Ramos-Hernanz, J.A.; Turkieh, M.J.; Badawi, M.; AlZoubi, H.; Abdus Sakur, M. Maximum Power Point Tracking of PV System Based on Machine Learning. *Energies* **2020**, *13*, 692. [CrossRef]

31. Du, Y.; Yan, K.; Ren, Z.; Xiao, W. Designing localized MPPT for PV systems using fuzzy-weighted extreme learning machine. *Energies* **2018**, *11*, 2615. [CrossRef]

32. Ahmad, S.; Hasan, N.; Bharath Kurukuru, V.S.; Ali Khan, M.; Haque, A. Fault Classification for Single Phase Photovoltaic Systems using Machine Learning Techniques. In Proceedings of the 2018 8th IEEE India International Conference on Power Electronics (IICPE), Jaipur, India, 13–15 December 2018; pp. 1–6.

33. Laudani, A.; Riganti Fulginei, F.; Salvini, A. High performing extraction procedure for the one-diode model of a photovoltaic panel from experimental I-V curves by using reduced forms. *Sol. Energy* **2014**, *103*, 316–326. [CrossRef]

34. Shiau, J.K.; Ma, C.W. Li-Ion battery charging with a buck-boost power converter for a solar powered battery management system. *Energies* **2013**, *6*, 1669–1699. [CrossRef]

35. Tang, M.; Stuart, T. Selective buck-boost equalizer for series battery packs. *IEEE Trans. Aerosp. Electron. Syst.* **2000**, *36*, 201–211. [CrossRef]

36. Arsalan, M.; Iftikhar, R.; Ahmad, I.; Hasan, A.; Sabahat, K.; Javeria, A. MPPT for photovoltaic system using nonlinear backstepping controller with integral action. *Sol. Energy* **2018**, *170*, 192–200. [CrossRef]

37. Davoudi, A.; Jatskevich, J.; De Rybel, T. Numerical state-space average-value modeling of PWM dc-dc converters operating in DCM and CCM. *IEEE Trans. Power Electron.* **2006**, *21*, 1003–1012. [CrossRef]

38. Pavlovic, T.; Bjazi, T.; Ban, Ž. Simplified averaged models of dc-dc power converters suitable for controller design and microgrid simulation. *IEEE Trans. Power Electron.* **2013**, *28*, 3266–3275. [CrossRef]

39. Babaei, E.; Seyed Mahmoodieh, M.E.; Mashinchi Mahery, H. Operational modes and output-voltage-ripple analysis and design considerations of buck-boost DC-DC converters. *IEEE Trans. Ind. Electron.* **2012**, *59*, 381–391. [CrossRef]

40. Texas Instruments SLVA721A—Basic Calculation of an Inverting Buck-Boost Power Stage. 2017. Available online: https://www.ti.com/lit/an/slva721a/slva721a.pdf (accessed on 29 September 2020).

41. Cheddadi, Y.; Errahimi, F.; Es-sbai, N. Design and verification of photovoltaic MPPT algorithm as an automotive-based embedded software. *Sol. Energy* **2018**, *171*, 414–425. [CrossRef]

42. Humada, A.M.; Hojabri, M.; Mekhilef, S.; Hamada, H.M. Solar cell parameters extraction based on single and double-diode models: A review. *Renew. Sustain. Energy Rev.* **2016**, *56*, 494–509. [CrossRef]

43. Mazhari, B. An improved solar cell circuit model for organic solar cells. *Sol. Energy Mater. Sol. Cells* **2006**, *90*, 1021–1033. [CrossRef]

44. Cheknane, A.; Hilal, H.S.; Djeffal, F.; Benyoucef, B.; Charles, J.-P. An equivalent circuit approach to organic solar cell modelling. *Microelectron. J.* **2008**, *39*, 1173–1180. [CrossRef]
45. Lun, S.; Wang, S.; Yang, G.H.; Guo, T.T. A new explicit double-diode modeling method based on Lambert W-function for photovoltaic arrays. *Sol. Energy* **2015**, *116*, 69–82. [CrossRef]
46. Gao, X.; Cui, Y.; Hu, J.; Xu, G.; Yu, Y. Lambert W-function based exact representation for double diode model of solar cells: Comparison on fitness and parameter extraction. *Energy Convers. Manag.* **2016**, *127*, 443–460. [CrossRef]
47. Wen, L.; Zhou, K.; Yang, S.; Lu, X. Optimal load dispatch of community microgrid with deep learning based solar power and load forecasting. *Energy* **2019**, *171*, 1053–1065. [CrossRef]
48. Li, L.; Ota, K.; Dong, M. When weather matters: IoT-based electrical load forecasting for smart grid. *IEEE Commun. Mag.* **2017**, *55*, 46–51. [CrossRef]
49. Raza, M.Q.; Khosravi, A. A review on artificial intelligence based load demand forecasting techniques for smart grid and buildings. *Renew. Sustain. Energy Rev.* **2015**, *50*, 1352–1372. [CrossRef]

Article

Power Scalable Bi-Directional DC-DC Conversion Solutions for Future Aircraft Applications

Antonio Lamantia [1], Francesco Giuliani [1] and Alberto Castellazzi [2,*]

[1]   Blu Electronic Srl, Via Lavoratori Autobianchi 1, 20033 Desio (MB), Italy; antonio.lamantia@bluelectronic.it (A.L.); fgiuliani121@gmail.com (F.G.)
[2]   Solid-State Power Processing (SP2) Lab, Faculty of Engineering, Kyoto University of Advanced Science, Kyoto 615-8577, Japan
*   Correspondence: alberto.castellazzi@kuas.ac.jp; Tel.: +81-75-496-6504

Received: 1 September 2020; Accepted: 9 October 2020; Published: 19 October 2020

Abstract: With the introduction of the more electric aircraft, there is growing emphasis on improving overall efficiency and thus gravimetric and volumetric power density, as well as smart functionalities and safety of an aircraft. In future on-board power distribution networks, so-called high voltage DC (HVDC, typically +/−270VDC) supplies will be introduced to facilitate distribution and reduce the associated mass and volume, including harness. Future aircraft power distribution systems will also very likely include energy storage devices (probably, batteries) for emergency back up and engine starting. Correspondingly, novel DC-DC conversion solutions are required, which can interface the traditional low voltage (28 V) DC bus with the new 270 V one. Such solutions presently need to cater for a significant degree of flexibility in their power ratings, power transfer capability and number of inputs/outputs. Specifically, multi-port power-scalable bi-directional converters are required. This paper presents the design and testing of such a solution, addressing the use of leading edge wide-band-gap (WBG) solid state technology, especially silicon carbide (SiC), for use as high-frequency switches within the bi-directional converter on the high-voltage side.

Keywords: DC-DC converters; multi-port dual-active bridge (DAB) converter; wide-band-gap (WBG) semiconductors; silicon carbide (SiC) MOSFETs; power converter

---

1. Introduction

The use of $115V_{AC}$ 400 Hz and $28V_{DC}$ power networks is a historical feature of avionic electrical power generation and distribution systems. The AC power is used directly for high power loads, such as starting, and is then rectified and conditioned to supply the bus power of $+28V_{DC}$ distributed to aircraft control systems, flight decks, and entertainment systems. However, there has been a pronounced movement within the aerospace industry to shift towards cleaner, more efficient and lower maintenance aircraft design as a result of the emergence of high fuel prices, the global warming problem, and high operating costs. The electrification of the aircraft to replace hydraulic or pneumatic functions with electrical ones is one of the prime movers in this field, as in the concepts of the *More* and *All* electric aircrafts [1–3]. This refers in particular to the replacement by electric actuators of complex aircraft hydraulic actuator systems, thereby dramatically reducing weight, maintenance costs, fuel consumption, footprint of carbon dioxide, and operating costs. Furthermore, the removal of pneumatic engine bleed systems makes it possible to run the engine more effectively and thus to save additional fuel.

The replacement of hydraulic and pneumatic systems with electric actuators and systems greatly increases the total electrical power requirements for an aircraft, requiring new approaches to the safe and intelligent delivery of aircraft power. To achieve the higher power ratings in a feasible way, novel enhanced aircraft Electrical Power Distribution System (EPDS) architectures investigated over recent years include smart power management systems, characterized by the presence of at least one high-voltage dc bus (HVDC, +/−270 V) and a low-voltage DC bus (LVDC, typically, 28 V), as illustrated very summarily, for instance, in Figure 1. Such architectures require the presence of HVDC/LVDC bi-directional DC-DC converters to interconnect and manage power transfer between the two bus levels, also enabling the addition of energy storage devices. Due to the abundance and variety of loads and still partly undefined power ratings (e.g., batteries) here, the focus is on a solution which enables a high degree of flexibility in relation to power scaling. In particular, a solution is pursued, in which a basic DC-DC converter power cell can be paralleled a number of times with control and supervision functionalities carried out by a unique central board. The prime drivers of the design are of course efficiency and power density, with an eye also to solutions meeting the single-fault tolerance expectations typical of avionic solutions. The design and test results of both single power cell and parallel operation are presented.

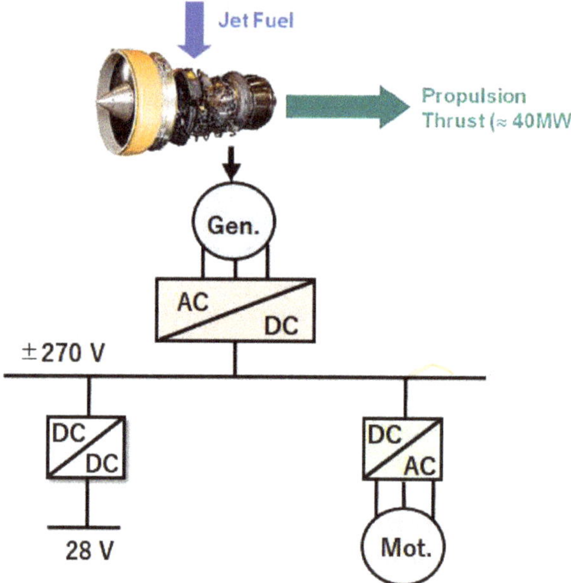

Figure 1. Generic illustration of future aircraft electrical network, including an high voltage dc (HVDC) power distribution bus: dedicated bi-directional DC-DC converters interface the 270 and 28 V buses; inverters cater for the AC loads (e.g., motors in pumps, actuators). Additional ports in the DC-DC converters may also be used to interconnect storage devices, with different voltage levels.

2. 270-28 V Bi-Directional Converter Design

The dual-active-bridge (DAB) topology was chosen among different solutions, for the following main advantages in this context [4,5]:

- 1st order dynamics and current generator equivalent characteristics, which enable straightforward parallel-ability and thus: (a) overall power scalability; (b) in-built redundancy and guarantee of reduced power operation capability; (c) overall efficiency optimization taking into account maximum efficiency versus load values of individual converter cells;
- easy extension to multi-port realization to interconnect, for instance, to storage devices [6];
- Zero-Voltage-Switching (ZVS) turn on of the power devices over a broad load range;
- suitability for use of planar magnetics design with possibility of integrated magnetics solutions;

- intrinsic current limitation in case of output short-circuit fault without the need for additional limiters or breakers.

DAB converter theory and operation is amply covered in literature already (see [2,3,6–8], for example) and so, here, the treatment of fundamental aspects is limited to some essential points of relevance to the subsequent discussion. For simplicity, we refer to the 270 V bus side as the input and the 28 V side as the output, but the discussion holds equal if the two are interchanged. The DAB consists of two full H-Bridges interconnected through a transformer, Figure 2a. The switches in diagonal pairs of either H-bridge are always turned on and off jointly and, within each bridge, each diagonal pair has 50% duty ratio. Control of output current delivery is achieved by introducing a phase-shift between the operation of the switches in the two H-bridges; with a series inductance all referred to the primary side, as in Figure 2a, the transformer secondary side gets directly connected to the output capacitor and so, the output voltage value is imposed back onto the primary scaled by the transformation ratio; the algebraic sum of input voltage and primary-reflected output voltage determines the voltage falling across the series inductor L_C during the phase-shift interval. Therefore, L_C can be effectively used as the current control element and its stored energy helps achieve turn-on soft switching of the power devices by means of resonance phenomena with the parasitic capacitance of the switches during the transitions. Both the power value and flow direction can be controlled by intervening on the value of the phase-shift Φ between the driving signals of the input and output H-bridges.

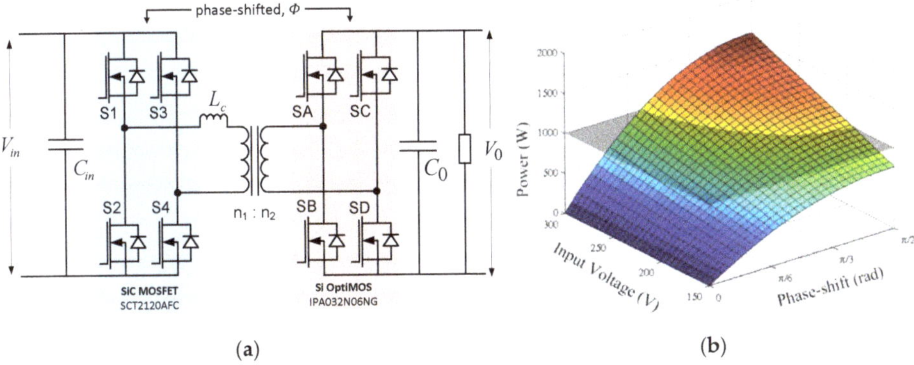

Figure 2. Main features of Dual-Active-Bridge converter: circuit schematic of the converter topology based on SiC MOSFETs or the primary and Si MOSFETs for the secondary side (**a**) and summary of output power delivery as a function of input voltage and phase-shift (**b**).

According to the fundamental model and neglecting losses, the power flow can be expressed by:

$$P_{DC} = \frac{V_{IN} \cdot V_O}{n \omega_s L_C} \cdot \Phi \left(1 - \frac{\Phi}{\pi} \right) \tag{1}$$

where V_{IN}, V_O are the values of the input and output DC voltages, n is the transformer turns ratio, $\omega_s = 2\pi f_s$ (f_s switching frequency), and Φ is the phase-shift (in radians). Equation (1) corresponds to a parabolic power delivery as a function of the phase-shift, with maximum power capability influenced by the value of input voltage, as summarized in Figure 2b considering a realistic variable input voltage range for the intended application.

2.1. Power Cell Design

Here, the intended nominal power rating of the *brick* power cell is 1.2 kW: more bricks can be connected in parallel to scale the power up. The primary (270 V) and secondary (28 V) side power cells are implemented using 650 V SiC MOSFETs (SCT2120AFC, ROHM Ltd., Kyoto, Japan) and 60 V

Opti-MOS (IPA032N06N3 G, Infineon Technologies, Munich, Germany), respectively. The choice of SiC devices for the primary (i.e., high voltage) side, as opposed to Si ones, is mainly motivated by the following reasons and advantages:

- Possibility to use a MOSFET type transistor as opposed to an IGBT, which allows for faster switching, removes the need for anti-parallel free-wheeling diode connection, and enables for more symmetrical operation in the forward and reverse power flow directions;
- Higher switching frequency capability with contained impact on efficiency, enabling the achievement of enhanced power density figures;
- Better temperature stability and higher temperature capability, which yield both more temperature independent efficiency levels over the intended operational range and favor long term reliability;
- Smaller overall intrinsic capacitances, which allow to achieve tun-on ZVS operation down to lower load values.

The gate drivers are bespoke designed, using non-symmetrical drive input voltage voltages for the SiC transistors on the primary side of −4 to +20 V and −4 to +15 V for the low voltage Si MOSFETs on the secondary side. Whereas a non-symmetrical drive voltage is a strict functional requirement of SiC MOSFETs, the choice to also apply an unsymmetrical driving voltage for the secondary-side Si transistors merely responds to cost containment targets, by enabling the design and production of a single gate-drive circuit, which can be adapted for both transistor types simply by replacing two ICs and one resistor in the circuit, with identical footprint. It should be noted moreover, that applying a somewhat higher (i.e., closer to zero) off-state bias voltage to the Si MOSFETs also allows to achieve faster switching transitions, with slightly reduced losses. Whereas the impact on overall efficiency is contained, the electro-thermal stress reduction in the transistor is interesting. For reference, the gate driver circuit schematic is shown in Figure 3. Insulation at gate-driver level is achieved by means of opto-couplers and signal transmission between control board and power cell is by means of optical fibers, ideal for avoiding electro-magnetic interference issues in reliability critical applications.

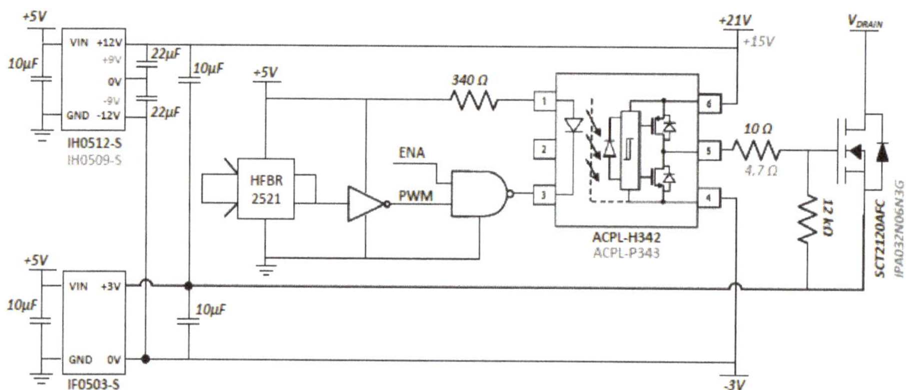

Figure 3. Schematic of gate driver circuit, featuring a unified design for both primary and secondary side transistors.

The magnetics are the other core components of the converter: the transformer provides galvanic isolation and voltage matching, the voltage across the series of its leakage inductance and the external inductor determines the current waveform flowing through the circuit. Since the transformer turns-ratio is relatively low, planar design of transformer and inductor are feasible and greatly interesting for increasing power density and improve thermal management; the components, shown in Figure 4, were custom designed and manufactured externally based on in-house specification. It is worth noting that the inductor could be realized as leakage inductance of the transformer, in the form of integrated magnetics. However, since its value is a key parameter of the converter control and dynamic characteristics, as well as of the ZVS load range, keeping the two components separate enables easier design optimization and better transformer thermal management. The switching frequency was chosen to be 100 kHz to start with and the transformer turns ratio was 19:2. After fixing the switching frequency and the turn ratio values, the maximum power and ZVS load range depend on the series inductor and the controllable phase-shift. Hence, an inductance value capable to satisfy the power constraints needs to been selected.

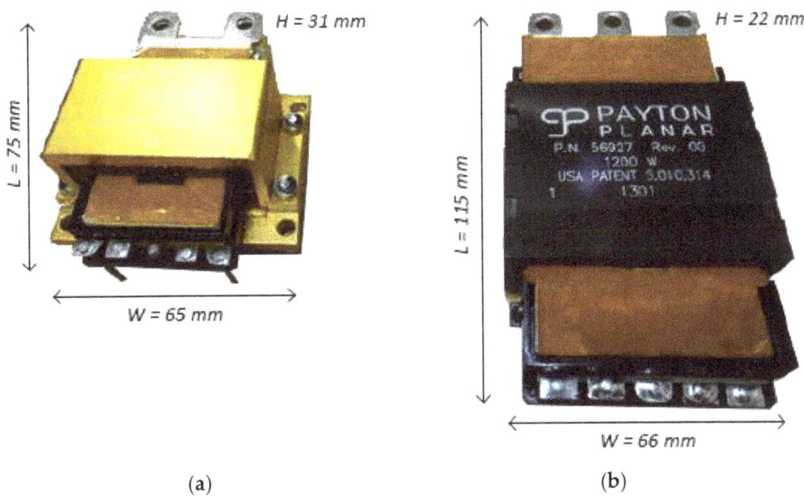

Figure 4. Photograph of planar magnetic components: heat-sinked inductor, (**a**), and transformer, (**b**).

2.2. Control Design

The goal of the control strategy is to keep the DC bus voltage constant at the desired voltage reference (28 V) in both forward and backward power flow, adjusting the phase-shift between the primary and secondary bridge. Starting from the fundamental reduced model of the DAB converter shown in Figure 5 along with the representative voltage and current waveforms illustrating basic operation and control of the converter, a state-space-average (SSA) model of the system was developed.

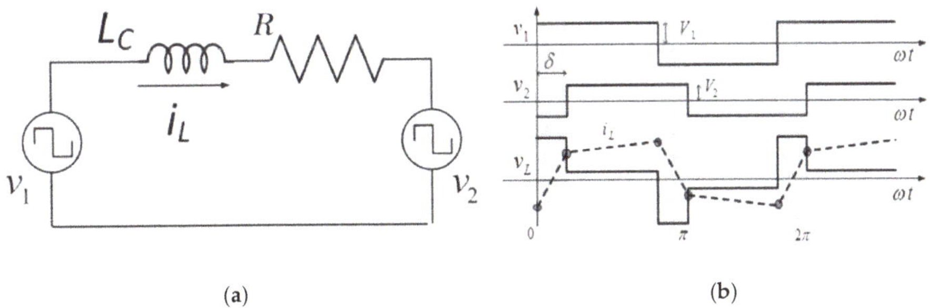

Figure 5. Dual-active bridge (DAB) basic equivalent state-space average model (**a**), and characteristic voltage and current waveforms illustrating the control principle (**b**).

The open loop dynamic characteristics of the converter are shown in Figure 6, in terms of its Bode-plot gain and frequency Bode-plots. It is important to underline that the converter DC gain, bandwidth and phase-margin are load-dependent, an aspect which needs to be duly taken into account when closing the loop.

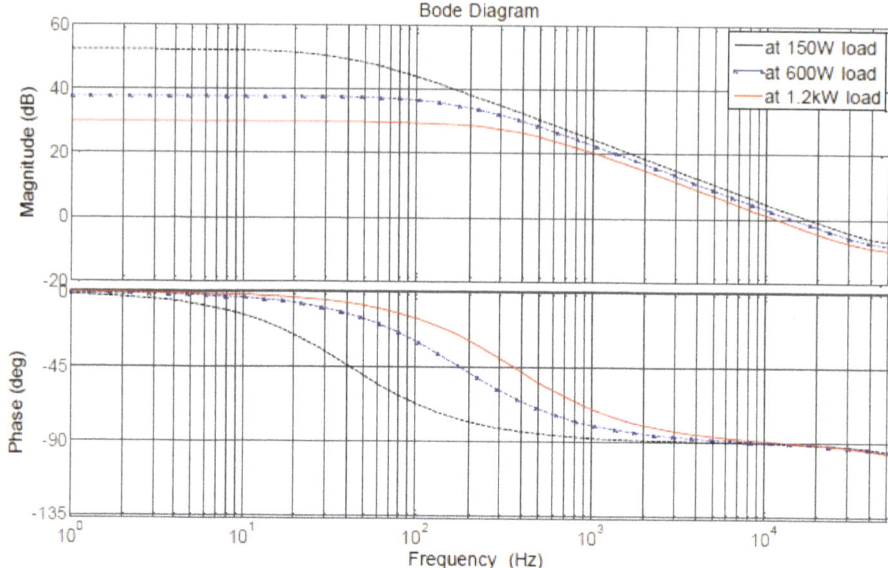

Figure 6. Open-loop Bode-plots of gain and phase for the power converter cell.

The chosen control system block diagram is shown in Figure 7. It consists of DC bus voltage control loop with a feed-forward current loop. The measured voltage V_O is compared with the reference voltage (28 V) and the error is fed through a PI controller to generate the desired phase-shift between the primary and secondary voltage square-waves. A feed-forward phase-shift compensation is added to make the system response faster. Based on the analytical model of the converter, the PI parameters are chosen to obtain the desired bandwidth with a proper phase margin. The phase-shift Φ depends non-linearly on load resistance R_O as:

$$\Phi = \frac{1 - \sqrt{1 - \frac{8nfL_cV_o}{V_iR_o}}}{2} = \frac{1 - \sqrt{1 - \frac{8nfL_cI_o}{V_i}}}{2} \tag{2}$$

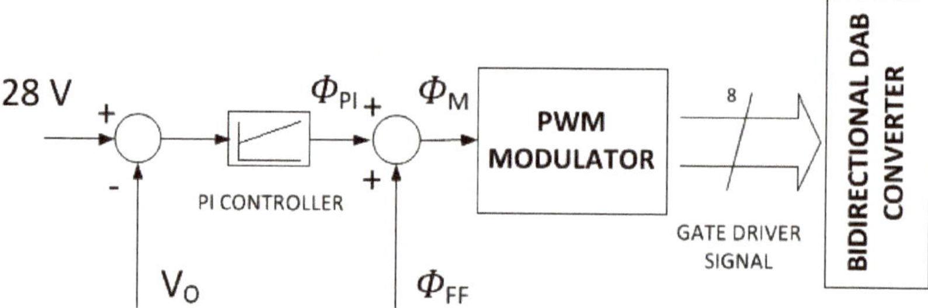

Figure 7. Converter control scheme with both a feedback and a feed-forward loop.

Since the switching frequency and the inductance are known, the output current and the input voltage are acquired by the sensors, it is possible to estimate the proper phase-shift during the working operations and let the PI controller manage only the small variations. Simulation results for the closed-loop response with the introduced control are summarized in Figure 8a, while in Figure 8b the converter step response for different start-up power levels is reported. It is clear that load dependence of the dynamics characteristics is significantly contained with the chosen approach.

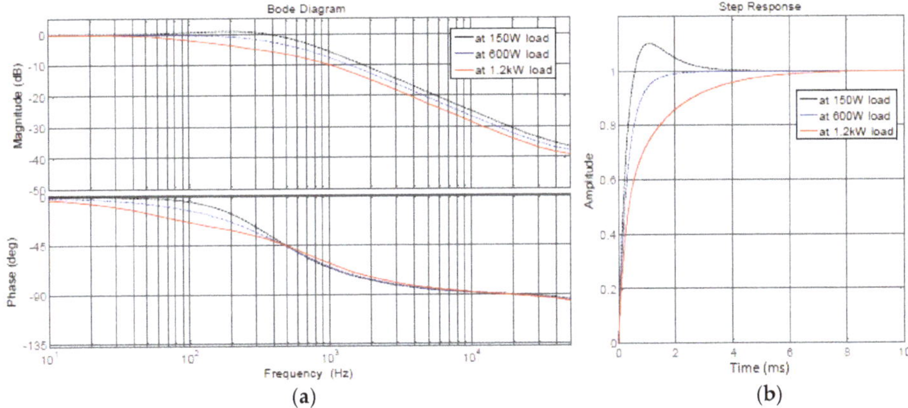

Figure 8. Dynamic response characterization of the converter (simulation results): closed-loop Bode plots (**a**) and system step response at different power levels (**b**).

The design phase also addressed EMI filtering on the 270 V side, ensuring compliance of standard avionic requirements (e.g., MIL-STD-704F) with a standard and compact damped LC filter design.

2.3. Experimental Characterisation

Figure 9 shows the engineered TRL6 level DC-DC converter, including control platform and flight-like enclosure, with details of connectors and wiring. Figure 10a shows some representative voltage and current waveforms at about 50% load (600 W) and Figure 10b summarizes measured efficiency data, still at 600 W, as a function of the input voltage in the range 230–290 V. Efficiency over load in the forward and backward power flow direction is reported in Figure 11: as can be seen, in the design and engineering of the converter, attention was taken to ensure high performance at relatively low loads; such approach is dictated by the actual mission profile of the converter, which foresees low-load operation as the most frequent condition, with the requirement to deliver the full power capability limited only to very short time durations. Therefore, it is very important to consider *energy* (as opposed to *power*) efficiency; that is, the integral over time of the power losses at the various load conditions to maximize overall aircraft efficiency. Moreover, it should be noted that due to the intended parallel operation of the converters, more units can be made to work at an optimized efficiency level even when the full capability of one single converter is not exhausted: for instance, if the total power demand is 900 W, it is more efficient to run tow cells in parallel than a single cell. The above represent a novel approach to design efficient power conversion systems, which will become increasingly important as the electrification level increases. It is also worth noting here, that bi-directional operation is mainly requested for enabling the possibility to use batteries connected on the lower-voltage side to be used as power source during some abnormal operational regimes or to implement regenerative energy storage functionalities. Therefore, the forward power transfer direction is still to be regarded as the primary one for the characterization for the converter.

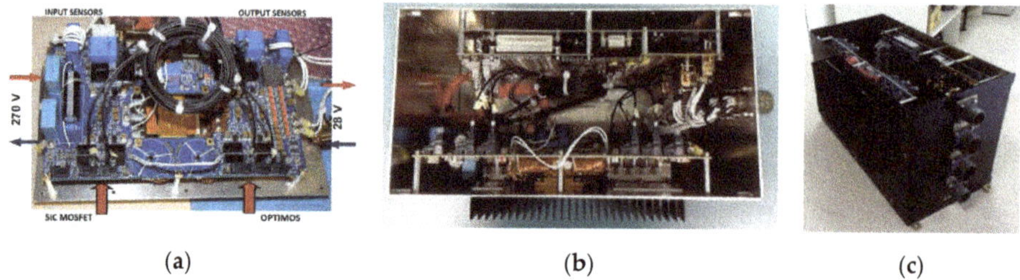

Figure 9. Photographs of developed hardware: single DC-DC converter cell (a); view of cased converter and control board, with details of wiring, and connectors (b); external view of cased unit (c).

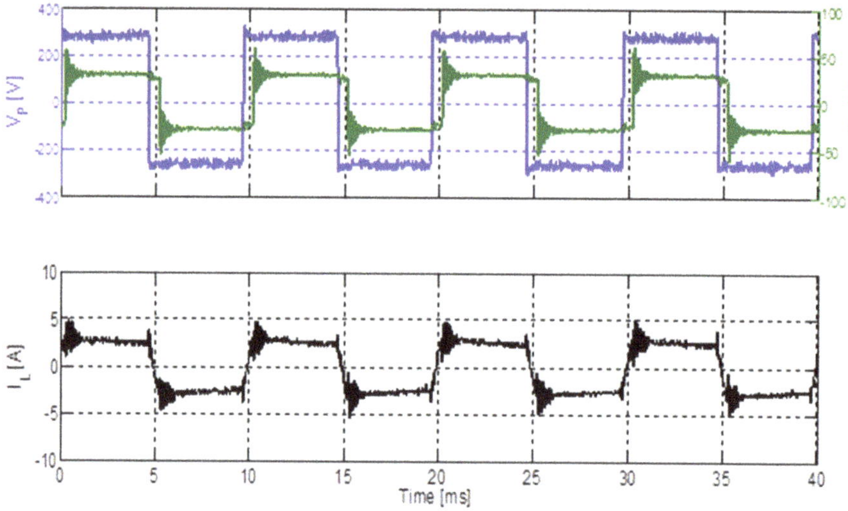

(a)

Efficiency measurement for variable input voltage							
V_{IN} (V)	I_{IN} (A)	P_{IN} (W)	V_{OUT} (V)	I_{OUT} (A)	P_{OUT} (W)	P_D (W)	η
235	2.8	653.0	28.1	21.2	597.1	55.86	91.4
240	2.7	650.0	28.1	21.3	597.8	52.20	92.0
250	2.6	657.5	28.2	21.6	608.6	48.91	92.6
260	2.5	650.8	28.2	21.5	607.3	43.48	93.3
270	2.4	645.6	28.3	21.5	606.5	39.01	94.0
280	2.3	641.3	28.3	21.4	605.6	35.78	94.4
290	2.2	638.2	28.3	21.4	604.5	33.74	94.7

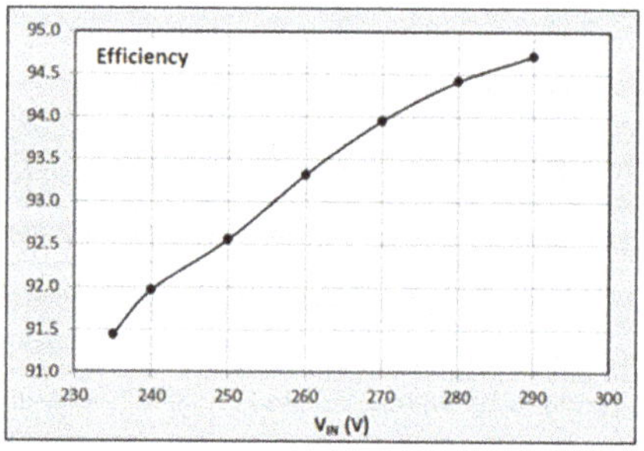

(b)

Figure 10. In (a), transformer primary and secondary voltage waveforms (top) and series inductor current (bottom); in (b), efficiency and detail of average voltage and current values.

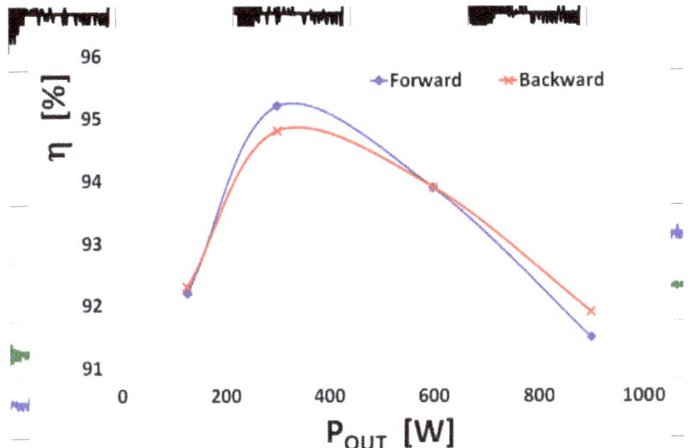

Figure 11. Measured converter efficiency in the forward and backward power transfer mode.

It should be noted that, although some ringing due to parasitic inductance is visible on the secondary side voltage waveform, no snubbers were eventually used in the converter implementation due to the Si MOSFETs being fully avalanche rated by design and not giving signs of any degradation over extended test periods for the parasitic energies involved in the switching transitions here. Indeed, the transistors are never driven into avalanche up to full load.

Some examples of closed-loop dynamic response are reported in Figure 12. Tests on the prototype confirmed the anticipated benefits of the chosen control approach, with consistent performance for both directions of power flow: the presence of the additional feed-forward loop is profitable and enables the converter to promptly manage large and rapid changes of power demand without unacceptable rising or falling of the voltage on the 28 V bus.

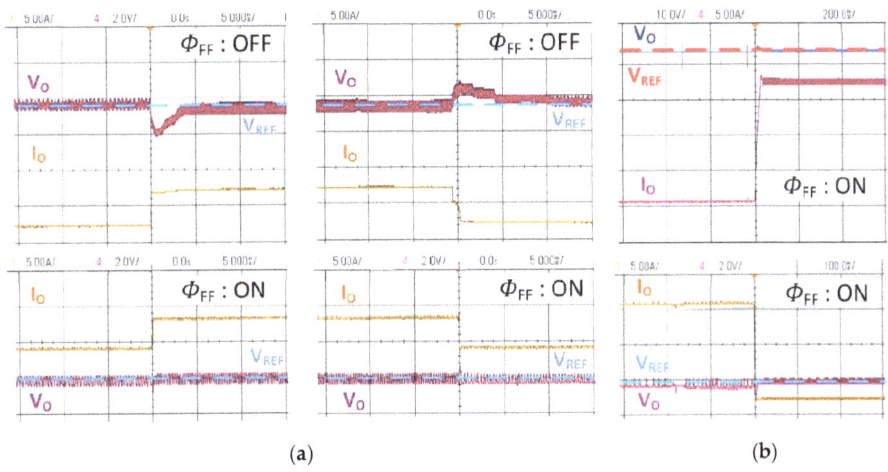

Figure 12. Scope snapshots of output voltage variation in the presence of abrupt and large step changes in the load (±5 A in (**a**) and ±15 A in (**b**)), without (top) and with (bottom) feed-forward loop, respectively.

3. Converter Paralleling for Power Scalability

To ensure a modular system architecture, with all its associated benefits, the parallel operation of more converters was further considered. The basic straightforward parallel connection of two converter units directly linked to the load is illustrated with the help of Figure 13. Due to its current source dynamic characteristics, paralleling of DAB converters is relatively straightforward and bi-directional converter (BDC) units are theoretically capable of working together without any additional control [9]. However, as soon as an unbalance arises, there is no way to equally split the power among the two modules without control. That is clearly shown in the results of Figure 14a,b. Moreover, a potential drawback is related to the bi-directional nature of the DAB converter: an issue could arise if the balance is established with one converter that works in regenerative configuration, as per experimental results of Figure 14c,d. In such a worst-case scenario, the paralleling of two units is not only pointless but even potentially destructive. For those reasons, a technique is needed for proper interconnections of BDC units.

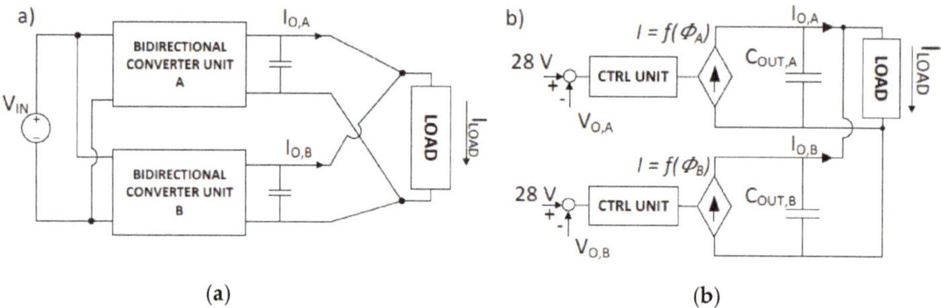

Figure 13. (**a**) Parallel connection of 2 BDCs; (**b**) controlled current source equivalent model.

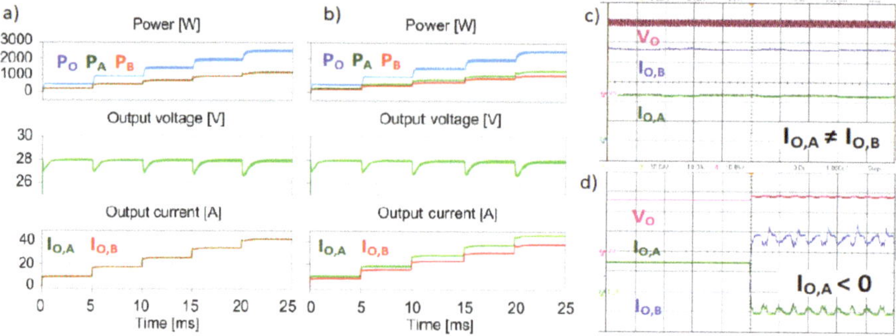

Figure 14. Parallel connection of BDC units without additional control: balanced, (**a**), and unbalanced, (**b**) conditions in simulation; balanced (**c**) and unbalanced (**d**) conditions in experimental tests.

Therefore, in this configuration, the aim of the control was still to regulate the load voltage, but also to ensure proper sharing of power delivery between the two modules. The corresponding equivalent control scheme in this case is illustrated in Figure 15. The methodology employed here consists in decoupling the control loops: one converter, acting as the master unit, ensures control of the output voltage to a constant value with the modality described in the previous section; the other, acting as a slave unit, is in charge of power sharing and operates in current control mode. The output current of each converter is sensed and the information is shared via a dedicated communication bus. The voltage controller in the salve unit is disabled and the master output current is fed-back as the set-point of the slave feedback-loop to determine the phase-shift. The feed-forward loop can also be added with a similar approach for both modules.

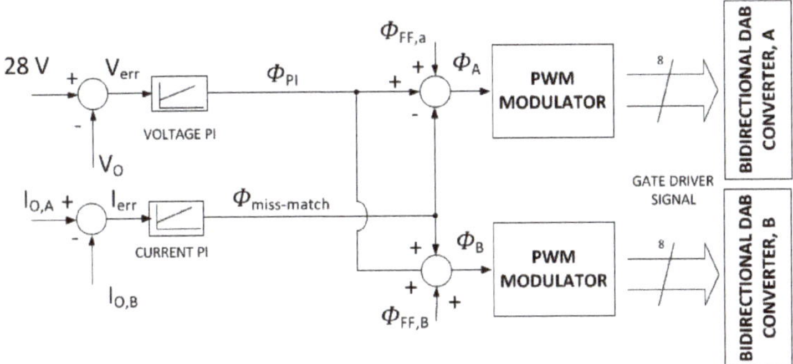

Figure 15. Structure of closed-loop system for parallel operation with current compensator.

Figure 16 shows the results obtained on the lab-prototypes for different power requests, after the implementation of this method. The tests confirm that the management of the energy is realized in good agreement with simulations and the output voltage is well regulated. The main advantage of the proposed solution is the fully controllable load sharing, which can be achieved with great accuracy in a simple and straightforward manner; the main limitation to its deployment is the need to exchange information between converters, which can become problematic in the case of multiple parallel modules if fast dynamic response is required. Loss of the communication-link or failure of the master unit would result in system shut-down.

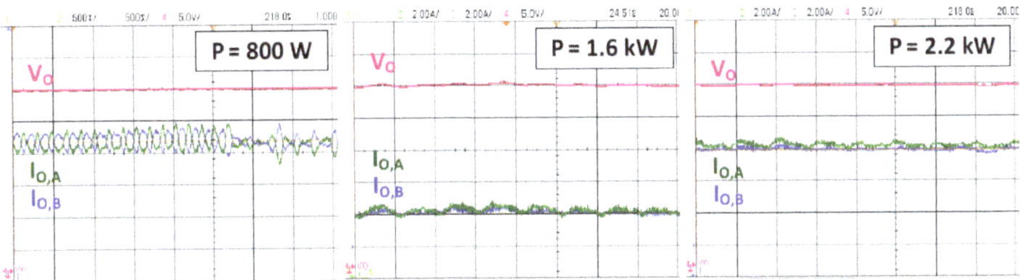

Figure 16. Experimental results under varying loads when using the current sharing parallel method.

Droop control is also a well-known practice used in power systems to share power among different generators. This concept has been recently proposed for other kinds of applications, such as speed control for integrated modular motor drives or modular DC/DC converter for smart transformers [10,11]. In this case, the droop control, illustrated in Figure 15, could be a suitable solution to enable parallel connection and to allow power sharing among the DABs; moreover, it does not require any communications between the BDC modules. The concept is to add a virtual resistor (K_{droop}) into the feedback loop, which drops the internal voltage set-point ($V_{O,ref}$) as a linear function of the output current (I_o)

$$V_{o,ref} = V^*_{o,ref} - K_{droop} \cdot I_o \tag{3}$$

Through this strategy, each converter exhibits a self-balancing characteristic, so it is not necessary for the single unit to be aware of the other elements in the system. The undesired effect of a steady-state error on the regulated voltage could be reduced or completely removed by another PI regulator, as also depicted in Figure 17.

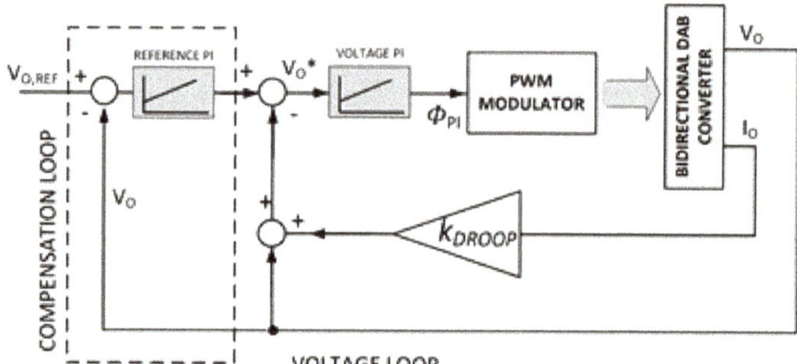

Figure 17. Block diagram illustration of the voltage-droop control with reference compensation protocol.

The results in Figure 18 show each converter working independently with the proposed voltage droop method, which intrinsically equalizes the power flow between the two BDC units connected in parallel to a resistive load. As the ratio of the droop coefficients is directly related to the amount of current flowing in each module, an asymmetrical power distribution is also possible simply adopting different virtual resistors for the single units (Figure 18b,c). In the system under analysis, the outer loop cut-off frequency was around one-third that of the inner loop.

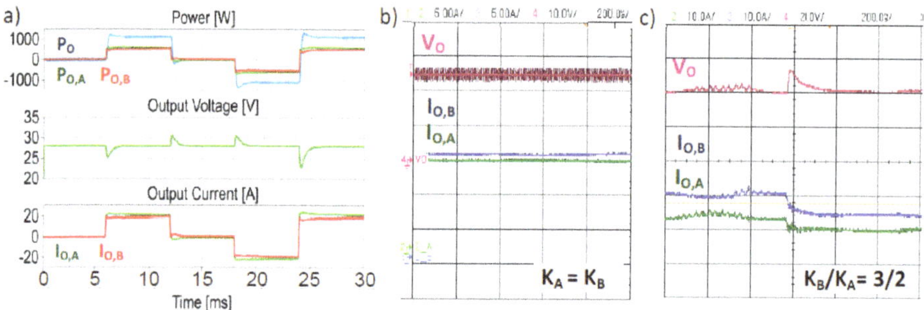

Figure 18. Parallel operation with droop control: converters working independently (**a**) and details of current unbalance between the two in the steady-state (**b**) and transient (**c**) regimes.

4. Conclusions

This paper has presented the design and development of a solution for implementing power-scalable bi-directional DC-DC conversion in future aircraft power networks [12,13]. The solution is based on the use of a dual-active bridge topology, which can easily be extended to multiple ports versions when the need to interface additional elements, such as batteries for storage, emerges [6,14]. In addition, the topology is suitable for the use of novel semiconductor technologies on the high-voltage side; in particular, silicon carbide (SiC) MOSFETs can yield important gains in switching frequency capability and converter performance optimization in terms of efficiency and thermal management requirements [14,15]. In the future, the possibility to develop bespoke modules will enhance the potential for disruptive progress in the integration level that can be realistically achieved [16].

Author Contributions: A.L. co-funded and managed the R&D activity and ensured technical lead in the engineering phase to TRL6 level; F.G. was the person in charge of the design and testing of the converter prototype and co-authored the paper; A.C. co-funded and supervised the research and development activity and co-authored the paper. All authors have read and agreed to the published version of the manuscript.

Funding: This research and development activity was partly funded by the EU FP7-JTI via the REGENESYS project (ID 308129).

Conflicts of Interest: The authors declare no conflict of interest.

References

1. Buticchi, G.; Bozhko, S.; Liserre, M.; Wheeler, P.; Al-Haddad, K. On-board Microgrids for the More Electric Aircraft—Technology Review. *IEEE Trans. Ind. Electron.* **2019**, *66*, 5588–5599. [CrossRef]
2. De, D.; Castellazzi, A.; Lamantia, A. 1.2kW dual-active bridge converter using SiC power MOSFETs and planar magnetics. In Proceedings of the 2014 International Power Electronics Conference (IPEC-Hiroshima 2014—ECCE ASIA), Hiroshima, Japan, 18–21 May 2014.
3. De, D.; Castellazzi, A.; Lopez-Arevalo, S.; Lamantia, A. SiC MOSFET based Avionic Power Supply. In Proceedings of the 7th IET International Conference on Power Electronics, Machines and Drives (PEMD 2014), Manchester, UK, 8–10 April 2014.
4. Giuliani, F.; Dipankar, D.; Delmonte, N.; Castellazzi, A.; Cova, P. Robust snubberless soft-switching power converter using SiC Power MOSFETs and bespoke thermal design. *Microelectron. Reliab.* **2014**, *54*, 1916–1920. [CrossRef]
5. Giuliani, F.; Delmonte, N.; Cova, P.; Costabeber, A.; Castellazzi, A. Soft-starting procedure for dual active bridge converter. In Proceedings of the 2015 IEEE 16th Workshop on Control and Modeling for Power Electronics (COMPEL), Vancouver, BC, Canada, 12–15 July 2015.
6. Castellazzi, A.; Gurpinar, E.; Wang, Z.; Hussein, A.; Garcia-Fernandez, P. Impact of Wide-Bandgap Technology on Renewable Energy and Smart-Grid Power Conversion Applications Including Storage. *Energies* **2019**, *12*, 4462. [CrossRef]
7. Krismer, F.; Kolar, J.W. Efficiency-Optimized High-Current Dual Active Bridge Converter for Automotive Applications. *IEEE Trans. Ind. Electron.* **2012**, *59*, 2745–2760. [CrossRef]
8. Maharana, S.; Mukherjee, S.; De, D.; Dash, A.; Castellazzi, A. Study of Dual Active Bridge with Modified Modulation Techniques for Reduced Link Current Peak Stress and Harmonics Losses in Magnetics. *IEEE Trans. Ind. Appl.* **2020**, *56*, 5035–5045. [CrossRef]
9. Higa, H.; Itoh, J. Zero voltage switching over entire load range and wide voltage variation of parallelly-connected dual-active-bridge converter using power-circulating operation. In Proceedings of the 2017 IEEE 3rd International Future Energy Electronics Conference and ECCE Asia (IFEEC 2017—ECCE Asia), Kaohsiung, Taiwan, 3–7 June 2017.
10. Nasirian, V.; Davoudi, A.; Lewis, F.; Guerrero, J.M. Distributed Adaptive Droop Control for DC Distribution Systems. *IEEE Trans. Energy Convers.* **2014**, *29*, 944–956. [CrossRef]
11. Rodrigues, W.A.; Oliveira, T.R.; Morais, L.M.F.; Rosa, A.H.R. Voltage and Power Balance Strategy without Communication for a Modular Solid State Transformer Based on Adaptive Droop Control. *Energies* **2018**, *18*, 1802. [CrossRef]
12. Sarlioglu, B.; Morris, C.T. More Electric Aircraft: Review, Challenges, and Opportunities for Commercial Transport Aircraft. *IEEE Trans. Transp. Electrif.* **2015**, *1*, 54–64. [CrossRef]
13. Wang, C.; Chen, J. Investigation on the Selection of Electric Power System Architecture for Future More Electric Aircraft. *IEEE Trans. Transp. Electrif.* **2018**, *2*, 563–576.
14. Wang, Z.; Castellazzi, A.; Saeed, S.; Navarro-Rodríguez, Á.; Garcia-Fernandez, P. Impact of SiC technology in a three-port active bridge converter for energy storage integrated solid state transformer applications. In Proceedings of the 2016 IEEE 4th Workshop on Wide Bandgap Power Devices and Applications (WiPDA), Fayetteville, AR, USA, 7–9 November 2016.
15. Wang, Z.; Castellazzi, A. Device loss model of a fully SiC based dual active bridge considering the effect of synchronous rectification and deadtime. In Proceedings of the 2017 IEEE Southern Power Electronics Conference (SPEC), Puerto Varas, Chile, 4–7 December 2017.
16. Castellazzi, A.; Fayyaz, A.; Gurpinar, E.; Hussein, A.; Li, J.; Mouawad, B. Multi-Chip SiC MOSFET Power Modules for Standard Manufacturing, Mounting and Cooling. In Proceedings of the 2018 International Power Electronics Conference (IPEC-Niigata 2018—ECCE Asia), Niigata, Japan, 20–24 May 2018.

Publisher's Note: MDPI stays neutral with regard to jurisdictional claims in published maps and institutional affiliations.

Article

A Tool for Evaluating the Performance of SiC-Based Bidirectional Battery Chargers for Automotive Applications

Giuseppe Aiello [1], Mario Cacciato [2], Francesco Gennaro [1], Santi Agatino Rizzo [2,*], Giuseppe Scarcella [2] and Giacomo Scelba [2]

[1] STMicroelectronics, 95129 Catania, Italy; giuseppe.aiello01@st.com (G.A.); francesco.gennaro@st.com (F.G.)
[2] Department of Electrical, Electronics and Computer Engineering, University of Catania, 95129 Catania, Italy; mario.cacciato@unict.it (M.C.); giuseppe.scarcella@unict.it (G.S.); giacomo.scelba@unict.it (G.S.)
* Correspondence: santi.rizzo@unict.it

Received: 5 November 2020; Accepted: 14 December 2020; Published: 20 December 2020

Abstract: In this paper, a procedure to simulate an electronic power converter for control design and optimization purposes is proposed. For the addressed application, the converter uses SiC-MOSFET technology in bidirectional battery chargers composed of two power stages. The first stage consists of a single-phase AC/DC power factor correction synchronous rectifier. The following stage is a DC/DC dual active bridge. The converter has been modulated using a phase-shift technique which is able to manage bidirectional power flows. The development of a model-based simulation approach is essential to simplify the different design phases. Moreover, it is also important for the final validation of the control algorithm. A suitable tool consisting of a system-level simulation environment has been adopted. The tool is based on a block diagram design method accomplished using the Simulink toolbox in MATLAB™.

Keywords: automotive; battery charger; circuit modelling; power electronics; SiC MOSFET

---

1. Introduction

Powertrain electrification of electric vehicles (EVs) and plug-in hybrid vehicles (PHVs) has gained the attention of governments, media and the public as a possible alternative mode of supplying power to transport vehicles due to its inherent efficiency advantages, e.g., less CO_2 emissions, in comparison with internal combustion engine vehicles [1,2]. EVs are key elements for the worldwide upgrade to sustainable energy systems. On the one hand, they directly affect the transition to environmentally friendly transportation. On the other, they are useful for compensating for the effects of dispersed generation based on renewable energy resources [3]. In more detail, when the power available from these generators surpasses the local load, it may be necessary to cut the exceeding power to avoid misoperation conditions, or worse yet, service continuity reductions. This limitation in green energy utilization can be overcome with EVs, since they involve an increment in the local load. Moreover, they can be used as energy storage systems which are able to mitigate fluctuations in primary energy resources and, more generally, are useful when coping with optimal power flow [4].

As a consequence of the diffusion of EVs and PHVs, an increasing number of connections to the public electrical grid of smart on- and off- board battery chargers has occurred [5]. These components are of fundamental importance for managing the energy flows between vehicles and the AC grid. Recently, a new EV operating mode, called Vehicle to Grid (V2G), was proposed [6]. V2G enables the use of the EVs as distributed large energy storage systems connected to the grid when parked [7]. The reward for providing ancillary services makes V2G economically convenient which, in turn, enables a wider diffusion of EVs, leading to environmental benefits [8]. On the other hand, the control strategy must take battery degradation into account [9].

Several bidirectional battery chargers (BBCs) for V2G have been already treated in the literature to investigate viable methods to achieve a compact, efficient and inexpensive solution. In [10,11], two designs of single-phase on-board BBCs were proposed, aiming to show the feasibility of reactive power support to the utility grid. In particular, [11] deals with the advantage of using wide band-gap semiconductor devices at high frequencies to reduce the current ripple by implementing both hardware and control solutions, similar to those adopted in converters for fuel cell power units [12]. In [13], a simple and functional BBC topology for stationary application was introduced. This topology was specifically designed to enhance the capabilities of a joint operation with an energy management system exploiting a storage stage in a residential environment. A literature analysis highlighted the fact that a key issue is to design and test a suitable control strategy. More specifically, the evaluation of the modulation, as well as of some features (e.g., current ripple and load step response), requires proper testing of the control strategy in dynamic conditions on small-time scales. On the other hand, appropriate long-timescale tests to evaluate the energy management capabilities of BBCs must be also be performed. In some works [10–14], the development of a feasible converter model was needed to fulfil the specifications through a proper system design, optimizing the structure of the control strategy as well as the correct setting of the parameters for the controllers.

As in many physical system designs, the use of advanced computer-aided design (CAD) systems is important at different project stages [15]. At the beginning, they enable component sizing verification; subsequently, they are useful for offline control validation with uP-based simulators [16], where they are very helpful when applying a user-friendly GUI based simulation interface [17]. Other solutions which are increasingly being adopted in the industry are powerful real-time emulation systems based on FPGA, that are widely used both in power converters [18] and electrical drives applications [19,20], and are particularly useful for the study and testing of dangerous situations, such as systems faults [21].

In this framework, a proper design using an advanced simulator model is proposed in this paper. It enables the evaluation of the feasibility and the performance of the converter using CAD. This approach makes it possible to validate the operation of the BBC in both V2G and Grid to Vehicle (G2V) operating modes. The main contribution of this paper is to propose a tool with which to optimize the BBC design before constructing the converter prototype. Additionally, the model of SiC MOSFET power devices was integrated to exploit their advantages in BBCs. Indeed, such an approach can useful for the optimal design of other converters in automotive applications. Finally, a mock-up was realized and tested, obtaining valuable results. In detail, the converter investigated was a 5-kW, single-phase BBC with two conversion stages: an active front end (AFE) PWM rectifier and a cascade-connected dual active bridge (DAB) with high-frequency isolation. Such an architecture was adopted for its bidirectional power flow, galvanic isolation, high efficiency in a wide operating range and reduced size and weight. The last features are due to the high switching frequency reached thanks to the use of SiC MOSFET power devices [22]. Every apparatus connected to the grid has to meet the power quality standards; therefore, the converter first stage also included power factor correction (PFC) capability. Figure 1 shows a flowchart of the overall tool.

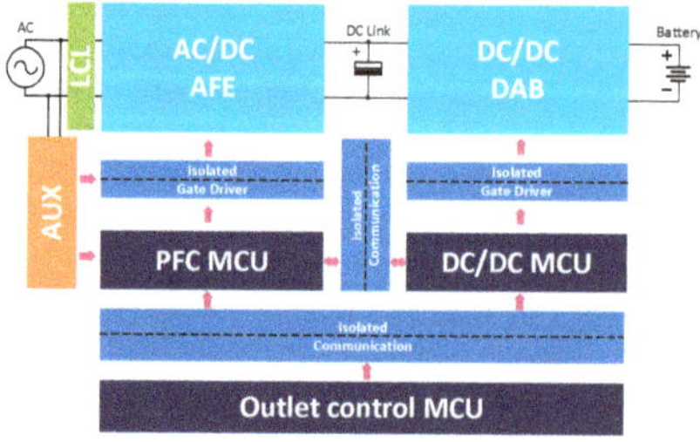

Figure 1. Overall picture of the evaluation tool.

2. Modelling the Bidirectional Battery Charger

For the application of a single-phase BBC, the proposed converter consists of two stages exploiting three H-bridges with modularity in the power board arrangement. Power devices with the same voltage breakdown should be used since the input and output voltage levels are similar. In this case, the three H-bridges can be identical, thus simplifying the converter design for the proposed converter that exploits identical SiC devices. As shown in Figure 2, the first stage is an AFE connected to the grid through an LCL filter which is useful to ensure both the power quality and the control of the power exchanged with the grid, while the second stage consists of a DAB converter.

The control strategy of the AC/DC converter is composed of a hierarchic control. On the one hand, it regulates the bidirectional power exchange with the grid. On the other, it shapes the current in a sinusoidal waveform. Hence, it consists of an inner loop current control in continuous conduction mode (CCM). The control is implemented on the dq rotating reference frame and is synchronous with the grid voltage. There is an outer loop to maintain constant the DC voltage, V_{DC}, using linear regulators, i.e., standard industrial proportional-integral (PI) control. As usual, the DAB is modulated in phase-shift. In this way, the control algorithm sets a suitable phase-shift for application between the switching signals of the two active bridges while maintaining the duty cycle of every switching pattern at 50%. Such a strategy makes it possible to achieve zero voltage switching (ZVS) upon turning on all of the DAB power switches, thereby increasing the converter efficiency. The phase-shift value sets the energy flow: in G2V mode, the energy flows towards the battery, while in V2G mode, the energy flow is directed from the battery to the AC grid.

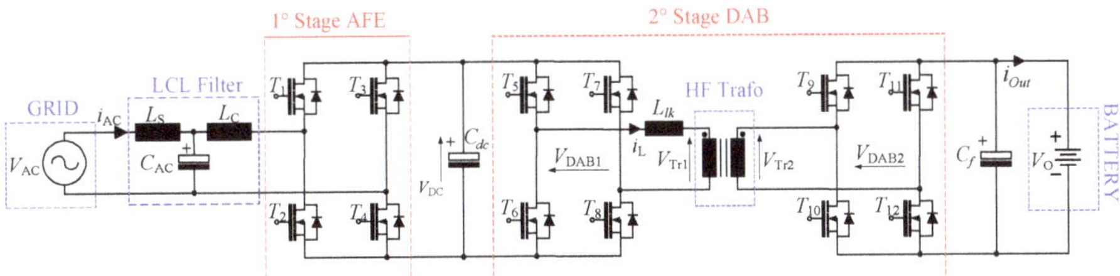

Figure 2. Converter topology. The first stage is the Active Front-End Rectifier; the second stage is the Dual Active Bridge.

2.1. Active Front-End Rectifier

The AFE, or synchronous rectifier, is connected via a filter to the utility grid where it performs AC/DC conversion and PFC [23,24]. Figure 3 shows the circuit test-bench emulator implementation using the Simulink Simscape Electrical Toolbox, a typical AFE control strategy based on the voltage oriented control algorithm. Park transformation is considered to obtain the best performance, e.g., zero error in steady-state and high control dynamics [25].

Park transformation is used to convert the two-phase stationary frame (α–β*) (1) into the two-phase rotating frame (d–q) which is synchronous with the grid voltage phase θ (2). The two-phase voltages in reference to the dq reference frame are converted in stationary quantities α–β using the inverse matrix of the reference frame transformation [26]:

$$\begin{cases} L\frac{di_\alpha}{dt} + Ri_\alpha = V_{t\alpha} - V_{s\alpha} \\ L\frac{di_\beta}{dt} + Ri_\beta = V_{t\beta} - V_{s\beta} \end{cases} \tag{1}$$

$$\begin{cases} L\frac{di_d}{dt} + Ri_d - \omega(t)Li_q = V_{td} - V_{sd} \\ L\frac{di_q}{dt} + Ri_q + \omega(t)Li_d = V_{tq} - V_{sq} \\ \qquad\quad \frac{d\rho}{dt} = \omega(t) \end{cases} \tag{2}$$

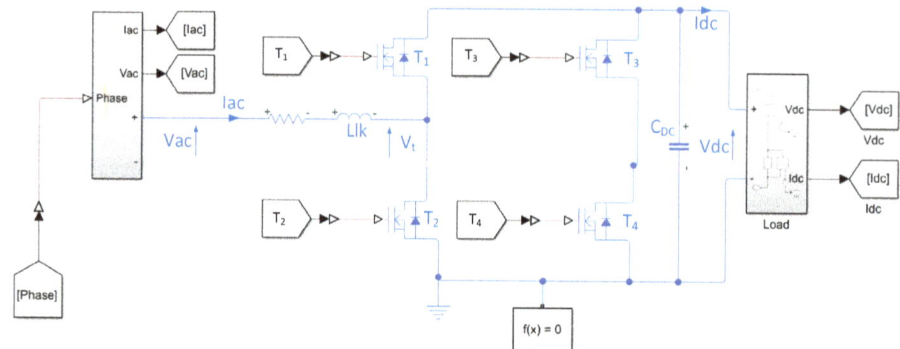

Figure 3. AC Stage—Implementation of the Active Rectifier in Simscape electrical Simulink.

The terms i_d, i_q and ρ are state variables, while V_{td}, V_{tq} and $\omega(t)$ are control variables. In particular, ω represents the control variable referring to the synchronous reference frame.

This model shows the nonlinearities related with the terms $\omega(t)Li_d$ and the sinusoidal components of the AC system: $V_{sd} = \hat{V}_S\cos(\omega_0 t + \theta_0 - \rho)$, $V_{sq} = \hat{V}_S\sin(\omega_0 t + \theta_0 - \rho)$.

Using this modelling approach, the purpose of the control is the cancellation of the sinusoidal terms; a phase locked loop "PLL" algorithm is used for this purpose. Applying the AC voltage as the input, the PLL output is $\rho(t) = \omega_0 t + \theta_0$, and Equation (2) turns to Equation (3), which contains only DC quantities in steady-state.

$$\begin{cases} L\dfrac{di_d}{dt} = \omega_0 Li_q - Ri_d + V_{td} - \hat{V}_S \\ L\dfrac{di_q}{dt} = -\omega_0 Li_d - Ri_q + V_{tq} \end{cases} \tag{3}$$

$$\begin{aligned} P_S(t) &= \frac{3}{2}\left[V_{sd}(t)i_d(t) + V_{sq}(t)i_q(t)\right] \\ Q_S(t) &= \frac{3}{2}\left[-V_{sd}(t)i_q(t) + V_{sq}(t)i_d(t)\right] \end{aligned} \tag{4}$$

$$\begin{aligned} P_S(t) &= \frac{3}{2}\left[V_{sd}(t)i_d(t)\right] \\ Q_S(t) &= -\frac{3}{2}\left[V_{sd}(t)i_q(t)\right] \end{aligned} \tag{5}$$

In V2G applications, the goal is to suitably manage the flow of active and reactive powers, according to Equation (4). By estimating the phase angle of the AC system through the PLL and imposing $V_{sq} = 0$, it follows that it is possible to rewrite the power relationships given in Equation (4) according to Equation (5), where the coupling terms have been cancelled.

Since in dq-axis, the component V_{sd} is constant, from Equation (5), it is evident that it is possible to obtain the power control $PQ_{ref} = PQ_{feed}$ through the direct control of the current $i_{dq\ ref} = i_{dq\ feed}$.

Finally, considering the general model, Equation (6), the command variables are obtained from the dq current control, Equation (7).

$$\begin{aligned} L\dfrac{di_d}{dt} &= \omega_0 Li_q - Ri_d + V_{td} - V_{sd} \\ L\dfrac{di_q}{dt} &= -\omega_0 Li_d - Ri_q + V_{tq} - V_{sq} \end{aligned} \tag{6}$$

$$V_{td} = u_d - \omega_0 Li_q + V_{sd}$$

$$V_{tq} = u_q + \omega_0 Li_d + V_{sq} \tag{7}$$

Simple and robust PI current regulators can be used to track references since the dq-axis signals in steady-state are constants (Equation (8)). The result is the controlled model given by Equations (9) and (10), as shown in Figure 4.

$$u_d = \left(k_p + \frac{k_i}{s}\right)\left(i_{d_{ref}} - i_{d_{feed}}\right)$$
$$u_q = \left(k_p + \frac{k_i}{s}\right)\left(i_{q_{ref}} - i_{q_{feed}}\right) \tag{8}$$

$$\begin{cases} V_{td} = \left(k_p + \frac{k_i}{s}\right)\left(i_{d_{ref}} - i_{d_{feed}}\right) - \omega_0 L i_q + V_{sd} \\[2mm] V_{tq} = \left(k_p + \frac{k_i}{s}\right)\left(i_{q_{ref}} - i_{q_{feed}}\right) + \omega_0 L i_q + V_{sq} \end{cases} \tag{9}$$

$$\begin{cases} L\dfrac{di_d}{dt} + R i_d - \omega L i_q = \left(k_p + \dfrac{k_i}{s}\right) \times \left(i_d^* - i_d\right) - \omega L i_q \\[3mm] L\dfrac{di_q}{dt} + R i_q + \omega L i_d = \left(k_p + \dfrac{k_i}{s}\right) \times \left(i_q^* - i_q\right) + \omega L i_q \end{cases} \tag{10}$$

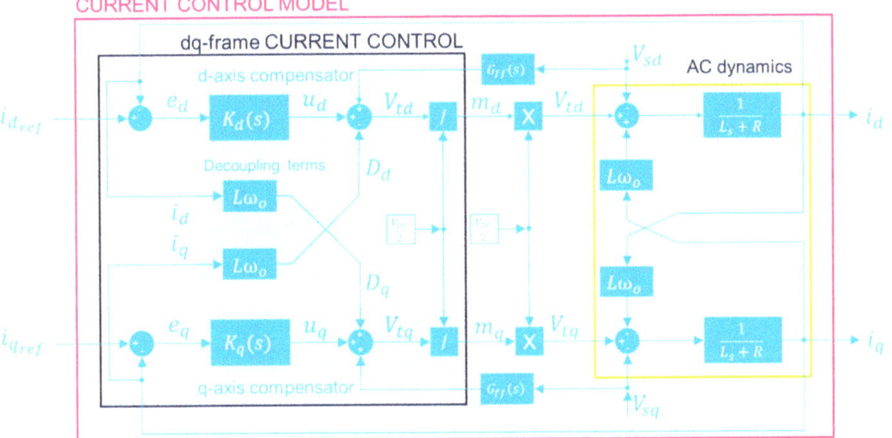

Figure 4. Current control diagram. $i_{dq_{ref}}$ current references, e_{dq} current errors, v_{dq} compensator terms, D_{dq} decoupling terms, $V_{t_{dq}}$ control voltages, m_{dq} modulation index.

2.2. Dual Active Bridge

The DAB is the DC/DC isolated bidirectional converter of the BBC (Figure 5). The DAB topology was chosen because of its high efficiency in a wide operating range [27]. It features a symmetrical structure, characterized by two full bridges connected via a high-frequency transformer which also provides galvanic isolation [28]. In Figure 6, a simplified equivalent circuit of the DAB converter is shown. The model of the transformer consists of two elements: the leakage inductor and an ideal transformer that models the voltage ratio. In Figure 7, a simplified circuit where the transformer has been represented on the secondary side to obtain a simple equivalent circuit is shown. The operating states of the converter switches are described in Equation (11).

$$v_{\text{DAB1}}(t) = \begin{cases} +V_1 & I & T_5, T_8 \ on \ \& \ T_6, T_7 \ off \\ 0 & II & T_5, T_7 \ on \ \& \ T_6, T_8 \ off \\ & III & T_5, T_7 \ off \ \& \ T_6, T_8 \ on \\ -V_1 & IV & T_6, T_7 \ on \ \& \ T_5, T_8 \ off \end{cases}$$
$$v_{\text{DAB2}}(t) = \begin{cases} +V_2 & I & T_9, T_{12} \ on \ \& \ T_{10}, T_{11} \ off \\ 0 & II & T_9, T_{11} \ on \ \& \ T_{10}, T_{12} \ off \\ & III & T_9, T_{11} \ off \ \& \ T_{10}, T_{12} \ on \\ -V_2 & IV & T_{10}, T_{11} \ on \ \& \ T_9, T_{12} \ off \end{cases} \tag{11}$$

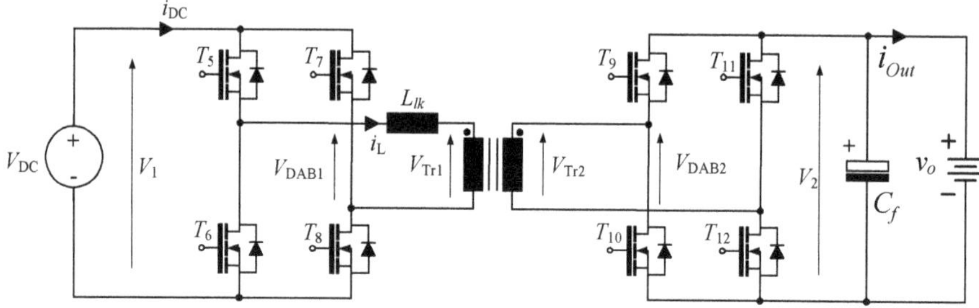

Figure 5. DC/DC stage Dual Active Bridge.

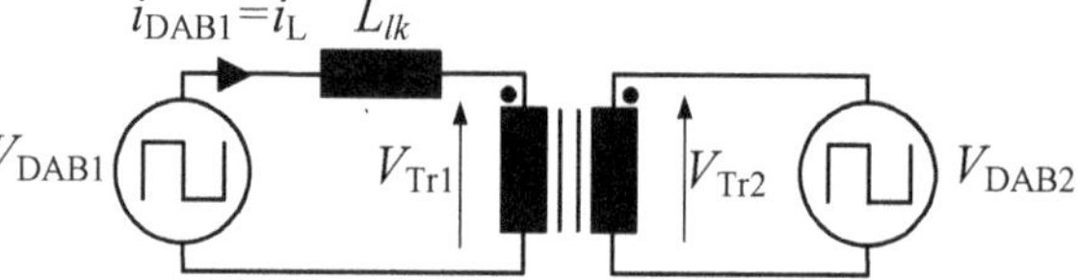

Figure 6. DAB equivalent circuit W/ transformer.

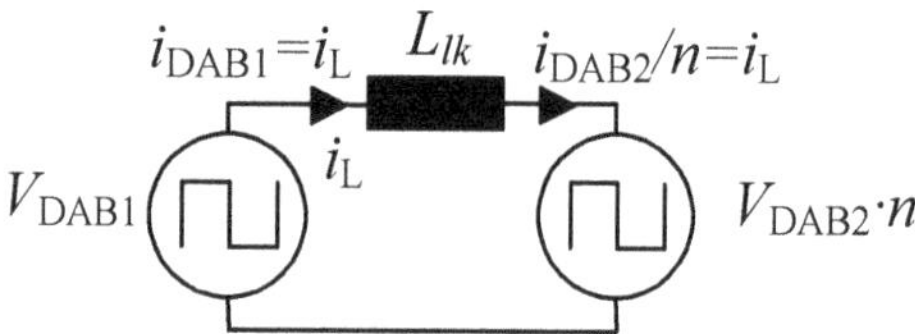

Figure 7. DAB equivalent circuit W/O transformer.

The H-Bridge on the left produces a square-wave voltage with a 50% of duty cycle on the primary side of the transformer. The right-side H-Bridge performs the AC to DC conversion and implements the current control loop used to shape the current charging profile of the battery. The leakage inductance, L_{lk}, plays a key role in the performance of the power conversion. Among the various modulation strategies suggested in the literature, single phase-shift modulation was used to control the power exchange between the BBC and the main grid.

The phase-shift (ϕ) is positive when the power flows from the grid to the battery and negative when the it flows in the opposite direction. The relation between the phase-shift and the delivered power is given by Equation (12):

$$P = P_{\text{DAB1}} = P_{\text{DAB2}} = \frac{nV_1V_2\phi(\pi - |\phi|)}{2\pi^2 f_s L_{lk}}, \quad -\pi < \phi < \pi \tag{12}$$

where $-180° < \phi < 180°$, V_1 and V_2 are the input and output voltages of the DAB (Figure 8); n is the transformer turn ratio; f_s is the switching frequency and L_{lk} is the leakage inductance when considering a lossless DAB model.

$P > 0$ denotes a power transfer from DAB1 to DAB2 and $P < 0$ denotes a power transfer from DAB1 to DAB2. The power transfer as a function of the phase-shift is depicted in Figure 8. The related absolute presents two maxima at two different phase-shift angles. The maximum power occurs for $\partial P/\partial \phi = 0$ is:

$$P|P_{max}| = \frac{n\,V_1\,V_2}{8\,f_s\,L_{lk}}, \quad \phi = \pm\frac{\pi}{2} \tag{13}$$

Hence, for a specific active power P, the phase-shift ϕ that must be imposed between the input-output voltages is:

$$\phi = \frac{\pi}{2}\left[1 - \sqrt{1 - \frac{8f_s L|P|}{nV_1 V_2}}\right]sgn(P) \tag{14}$$

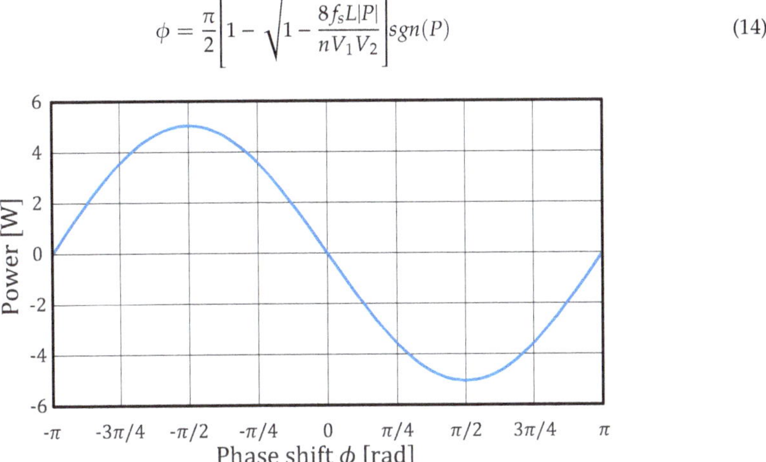

Figure 8. Power transfer vs. Phase-shift.

2.3. High-Frequency Transformer

The high-frequency transformer is responsible for the power transfer and permits to obtain the galvanic isolation [29]. Different core geometries and materials are widespread and the selection of the most appropriate solution mainly depends on the specific application. It is well known that the use of high switching frequency reduces the core size for a given power, while using suitable ferrite materials effectively eliminates eddy currents losses.

The design method is based on the "core geometry method" [30,31].

2.4. Matlab—Simulink Implementation

The model of the BBC and the model of its control were implemented in Matlab-Simulink to simulate the BBC behavior and to evaluate its performance considering different working conditions. The converter model included parasitic elements that affect each power conversion stage. The MOSFETs parameters were considered, as well as the dead-time set in the driving circuit. The closed-loop control block diagram for the AC/DC PFC converter is shown in Figure 9.

Using the Park's transformation, the regulation was implemented using the i_d and i_q current components to control, respectively, the active and reactive power. This control structure makes it possible to regulate both the DC voltage value and the PF. During G2V mode, the AFE with the PF correction works as an AC to DC converter, and charges the battery while maintaining constant the DC voltage and unitary the PF.

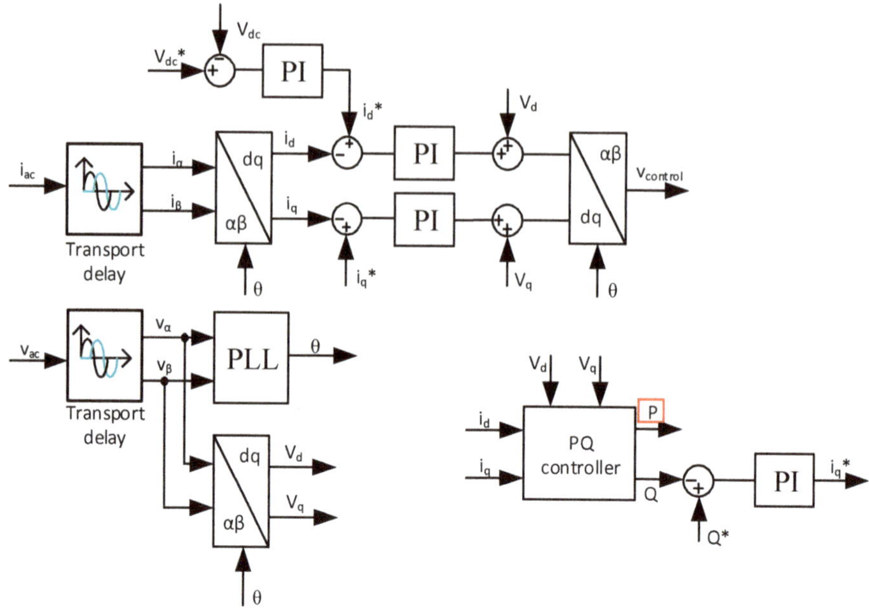

Figure 9. AFE PFC closed-loop control block diagram.

In V2G mode, the battery is discharged and the bridge acts in inverter mode (DC/AC). The control strategy consists of maintaining constant the voltage value on the bus-dc and managing the PF to compensate the amount of reactive power required by the grid. The control loop block diagram for the DAB is shown in Figure 10.

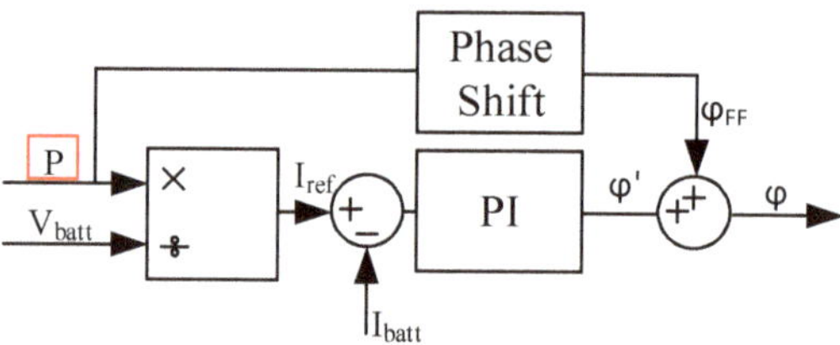

Figure 10. Phase-shift control loop block diagram.

3. Simulation and Validation of the Model of Bidirectional SiC-Based Battery Chargers

A bipolar PWM was implemented with a switching frequency f_s = 100 kHz. The switching frequency was selected as the best compromise between efficiency and high power density due to the reduction of the passives composing the AC grid filter and the DC bus link. The modulating signal was evaluated by the voltage grid angle implementing a grid synchronization algorithm setting a unity Power Factor (PF) in G2V or a stable grid synchronization in V2G. The gate signals used to control the SiC MOSFETs were set by the current control loop.

The technical specification of the filter parameters, DC bus link and grid operating conditions considered in the following analysis are listed in Table 1. The design specifications of the DAB of the proposed BBC are listed in Table 2.

Table 1. Technical specification of the AFE parameters.

Parameter	Value
RMS voltage grid	230 V
Grid frequency f_e	50 Hz
L_s	1.5 μH
Filter parameter C_{ac}	10 μF
Filter parameter L_c	325 μH
C_{dc}	400 μF
Switching frequency f_s	100 kHz

Table 2. DAB design specifications.

Parameter	Value
Nominal input Voltage V_{dc}	400 V
Nominal output voltage V_o	400 V
Minimal output voltage $V_{o,min}$	150 V
Output Power	5 kW
Duty Cycle	0.5
Switching frequency	100 kHz

For this bidirectional converter, the EE core geometry was chosen with N87 material grade. This choice was related to the high switching frequency (f_s = 100 kHz) and high-power density of the transformer, whose characteristics are listed in Table 3. An increment in the switching frequency enabled a reduction of passive component size and weight but at the cost of greater switching power losses that, in turn, involve reduced efficiency. Therefore, the adopted frequency was the best compromise for such an application.

Table 3. Technical specification of the transformer parameters.

Parameter	Value
Nominal Input Voltage	400 V
Maximum Input Voltage	480 V
Minimum Input Voltage	360 V
Input current	22 A
Nominal output voltage	400 V
Output Current	17.5 A
Regulation α	0.15%
Max operating flux density B_m	0.16 T
Maximum temperature rise T_r	70 °C

A prototype of the converter was designed and realized using components made by STMicroelectronics to validate the proposed tool by testing the performance and efficiency of the BBC designed using the proposed modelling approach. The power devices are SiC MOSFETs SCT50N120 (Table 4).

Table 4. Power device description: SiC MOSFET SCT50N120.

Symbol	Parameter	Value
V_{DS}	Maximum drain-source voltage	1200 V
Id	Drain current (continuous) at TC = 25 °C	65 A
R_{DS} (on)	Static drain-source on-resistance at 150 °C	59 mΩ
Tj	Max Operating junction temperature in HiP247™	200 °C

A mixed-signal MCUs STM32G474 was used to generate the phase-shift control signal and to manage the dead-time in each power converter leg exploiting the High-Resolution Timer (HRTIM) with

184 ps resolution. The digital control signals were conditioned and applied to the power switches using high-performance gate drivers STGAP2S, a galvanically isolated 4 A single gate drivers. This made it possible to achieve more compact and robust solutions for the entire experimental system.

The modular prototype and the test-bench are shown in Figures 11 and 12.

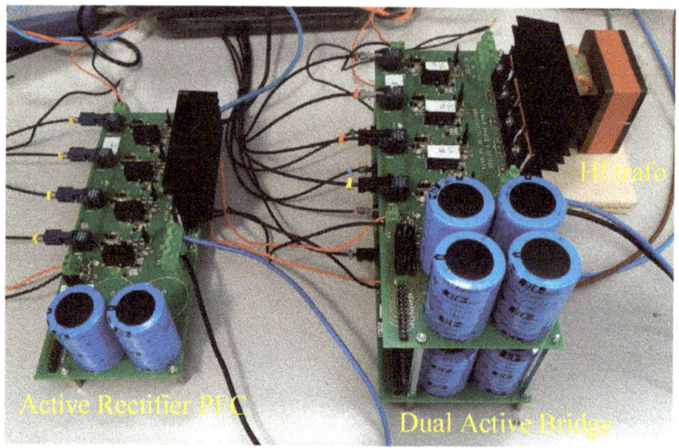

Figure 11. A prototype of the bidirectional battery charger.

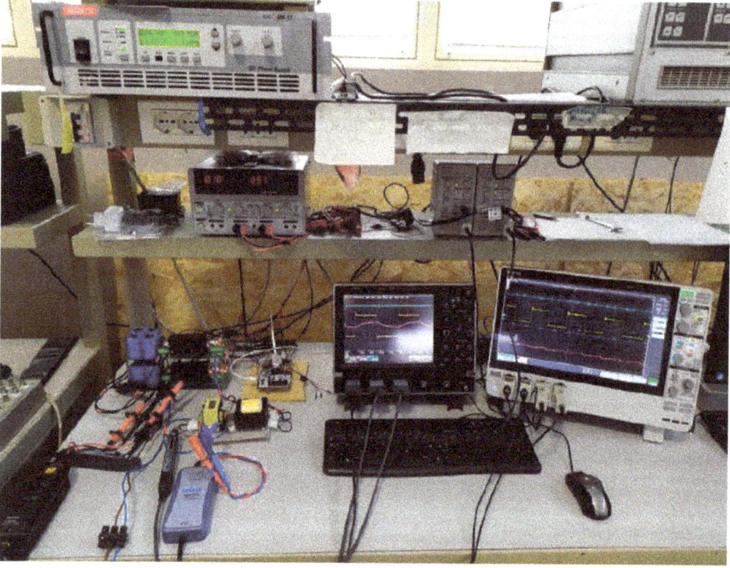

Figure 12. Prototype test-bench.

The power required from the AFE acts on the phase-shift; by varying this reference, it is possible to reverse the power flow. Some simulated and measured waveforms obtained during the G2V mode are shown below. In Figure 13, the simulated first stage waveforms that are the grid voltage v_{ac} and current i_{ac} with unitary PF, and the ripple of the DC voltage are shown. The total harmonic distortion for the AC current was close to 7%, which is in accordance with the value measured (less than 10%) using the prototype.

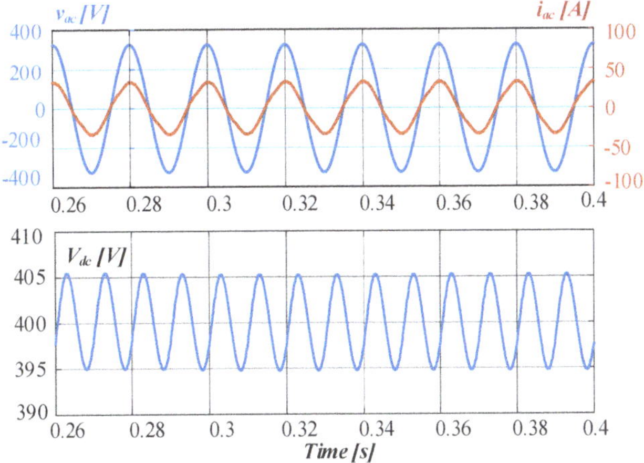

Figure 13. Voltage and current of the AC grid and ripple on the DC side.

The voltage and current on the primary side of the transformer are shown in Figure 14. The simulated waveforms were in good agreement with the measured ones. The main difference was the lack of oscillations in the simulated voltage. These oscillations were due to the coupling between the parasitic capacitance of the devices and the parasitic inductances in the power loop that were neglected in the model. The current waveform depends on the phase-shift between the two transformer-ends voltages. The secondary side quantities were pretty similar, as a turn ratio n equal to one was chosen. The leakage inductance, L_{lk}, affected the power delivered in the DAB converter. Therefore, the voltage v_L waveform was strictly related to the power direction. The DC output waveforms are shown in Figure 15, where the ripple of the voltage V_o and current I_o are highlighted.

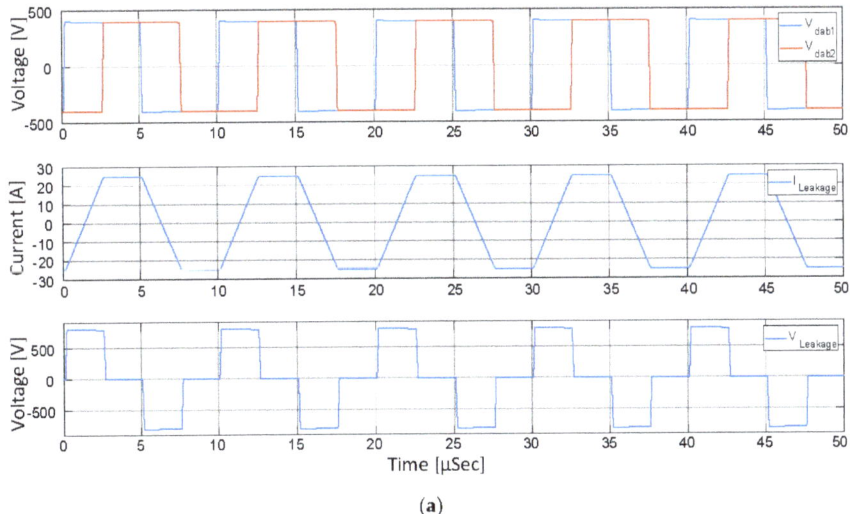

(**a**)

Figure 14. *Cont.*

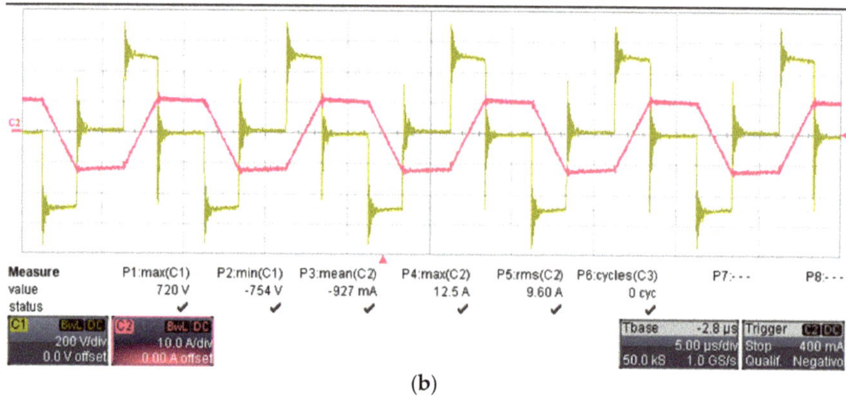

(b)

Figure 14. Voltage on both sides of the transformer, voltage and current of the inductance. (**a**) Simulated waveforms; (**b**) measured waveforms.

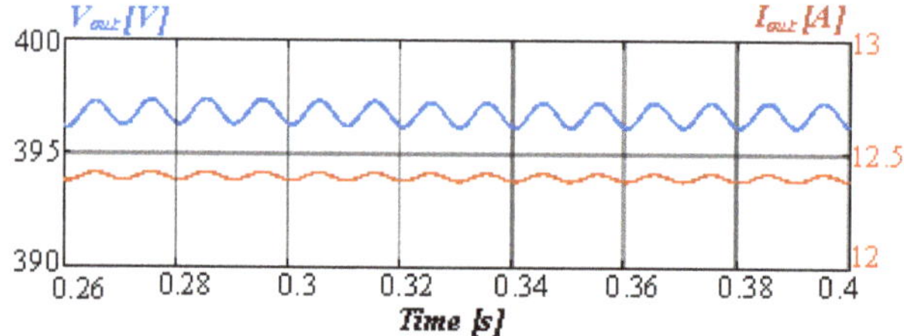

Figure 15. Ripple on the DC output Voltage and Current.

In V2G mode, the power flows from the battery to the grid to satisfy the power demand. In this case, the reference power is modified and acts on the phase-shift value as described above. The transition from G2V to V2G mode at the instant t* requires current inversion, as illustrated in Figure 16. In this case, the PF has been maintained, meaning that no reactive power was requested by the converter thanks to the proper control. The main simulation results are summarized in Table 5, while in Figure 17, the efficiency of the whole converter is shown.

Table 5. V2G operation—Simulation quantities and results.

Parameter	Value
RMS grid voltage V_s	230 V
RMS grid current I_s	22.6 A
Average Bus DC Voltage V_{dc}	403 V
Average output voltage V_o	397 V
Average output current I_o	12.4 A
Input Apparent Power S	5200 VA
Input Active Power P_{ac}	5200 W
Bus DC Power P_{dc}	5030 W
Output Power P_o	4910 W
Power Factor	0.999
Displacement Power Factor	1
Total Harmonic Distortion	7%
AFE efficiency $\eta = P_{DC}/P_{ac}$	96.7%
DAB efficiency $\eta = P_o/P_{DC}$	97.6%
Power Efficiency $\eta_p = P_o/P_{ac}$	94.42%
Conversion Factor $\eta_c = P_o/S$	94.26%

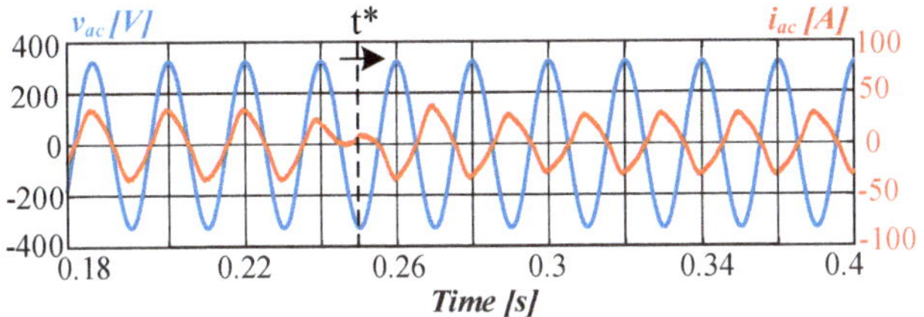

Figure 16. AC voltage and current from G2V to V2G mode.

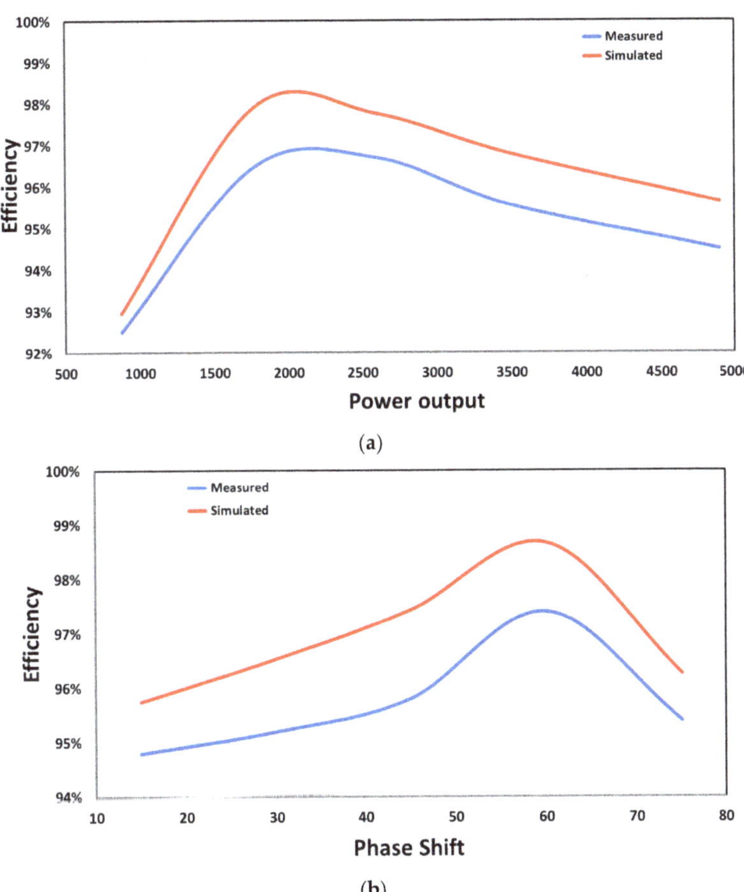

(a)

(b)

Figure 17. Efficiency vs. (**a**) power output and vs. (**b**) phase shift.

Some other comparisons are reported in Figure 18, confirming the consistency of the proposed modelling approach.

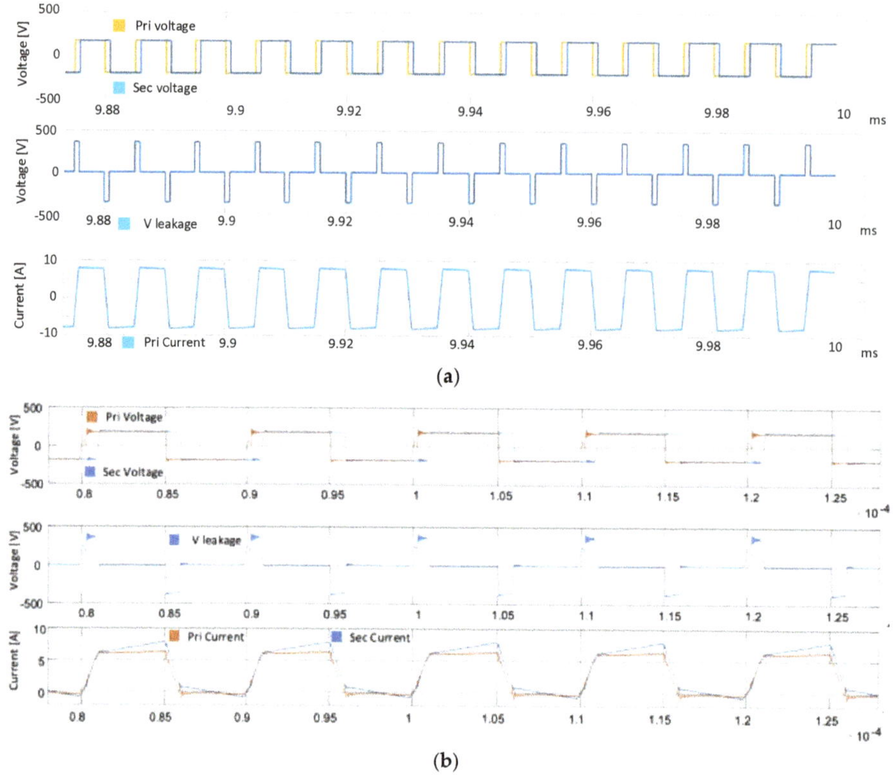

Figure 18. DAB: voltage and current waveforms. (**a**) Simulated (**b**) Measured.

4. Conclusions

This paper dealt with SiC MOSFET-based BBC with galvanic isolation. A promising topology was studied as the best choice in terms of efficiency, bidirectional power flow management and complexity. The development of an accurate tool accounting for the model of the converter in computer simulator and which was able to exploit FPGA was proposed. It has been shown that this is a suitable approach to design and test the performance of the complex control algorithm, both in G2V with PFC capability and V2G operation modes. The control strategy of the AC/DC converter is composed of a cascade control. One is able to regulate the power flow with the grid, while control of the DC/DC stage consists of the management of the battery charge/discharge. The design and the proposed approach were validated by comparing the simulation results with some experimental tests, confirming the consistency of the proposed method.

Author Contributions: Writing-Review & Editing: G.A., M.C., F.G., S.A.R., G.S. (Giuseppe Scarcella), G.S. (Giacomo Scelba). All authors have read and agreed to the published version of the manuscript.

Funding: This research received no external funding.

Conflicts of Interest: The authors declare no conflict of interest.

Nomenclature

AFE	Active Front End
BBC	Bidirectional Battery Charger
CCM	Continuous Conduction Mode

DAB	Dual Active Bridge
EV	Electric Vehicle
FPGA	Field Programmable Gate Array
GUI	Graphical user interface
HRTIM	High-Resolution Timer
PF	Power Factor
PFC	Power Factor Correction
PHV	Plug-in Hybrid Vehicle
PLL	Phase Locked Loop
SiC	Silicon Carbide
V2G	Vehicle to Grid
ZVS	Zero Voltage Switching
C_{ac}	LCL filter capacitor
C_{dc}	capacitor between the AFE and the DAB
C_f	output capacitor
L_C, L_S	LCL filter inductors
L_{lk}	transformer leakage inductor
i_α, i_β	two-phase stationary currents
i_d, i_q	two-phase rotating currents
i_{AC}	line current drawn by the converter
i_{Llk}	current flowing through the transformer leakage inductor
i_{out}	output current
v_{out}	output voltage
v_1	DC/DC input voltage
v_2	DC/DC output voltage
$v_{t\alpha}, v_{t\beta}$	two-phase stationary converter voltages
v_{td}, v_{tq}	two-phase rotating converter voltages
$v_{s\alpha}, v_{s\beta}$	two-phase stationary AC main voltages
v_{sd}, v_{sq}	two-phase rotating AC main voltages

References

1. Shi, J.; Gao, Y.; Wang, W.; Yu, N.; Ioannou, P.A. Operating Electric Vehicle Fleet for Ride-Hailing Services with Reinforcement Learning. *IEEE Trans. Intell. Transp. Syst.* **2020**, *21*, 4822–4834. [CrossRef]
2. Neffati, A.; Marzouki, A. Local energy management in hybrid electrical vehicle via Fuzzy rules system. *AIMS Energy* **2020**, *8*, 421–437. [CrossRef]
3. Dominkovic, D.F.; Bačeković, I.; Pedersen, A.S.; Krajačić, G. The future of transportation in sustainable energy systems: Opportunities and barriers in a clean energy transition. *Renew. Sustain. Energy Rev.* **2018**, *82*, 1823–1838. [CrossRef]
4. Franco, F.L.; Ricco, M.; Mandrioli, R.; Grandi, G. Electric Vehicle Aggregate Power Flow Prediction and Smart Charging System for Distributed Renewable Energy Self-Consumption Optimization. *Energies* **2020**, *13*, 5003. [CrossRef]
5. Vadi, S.; Bayindir, R.; Colak, A.M.; Hossain, E. Vadi A Review on Communication Standards and Charging Topologies of V2G and V2H Operation Strategies. *Energies* **2019**, *12*, 3748. [CrossRef]
6. Salvatti, G.A.; Carati, E.G.; Cardoso, R.; Da Costa, J.P.; Stein, C.M.D.O. Electric Vehicles Energy Management with V2G/G2V Multifactor Optimization of Smart Grids. *Energies* **2020**, *13*, 1191. [CrossRef]
7. Juul, F.; Negrete-Pincetic, M.; Macdonald, J.; Callaway, D. Real-time scheduling of electric vehicles for ancillary services. In Proceedings of the 2015 IEEE Power & Energy Society General Meeting, Denver, CO, USA, 26–30 July 2015; pp. 1–5.
8. Liu, C.; Chau, K.T.; Wu, D.; Gao, S. Opportunities and Challenges of Vehicle-to-Home, Vehicle-to-Vehicle, and Vehicle-to-Grid Technologies. *Proc. IEEE* **2013**, *101*, 2409–2427. [CrossRef]
9. Kisacikoglu, M.C.; Ozpineci, B.; Tolbert, L.M. EV/PHEV Bidirectional Charger Assessment for V2G Reactive Power Operation. *IEEE Trans. Power Electron.* **2013**, *28*, 5717–5727. [CrossRef]

10. Kisacikoglu, M.C.; Kesler, M.; Tolbert, L.M. Single-Phase On-Board Bidirectional PEV Charger for V2G Reactive Power Operation. *IEEE Trans. Smart Grid* **2015**, *6*, 767–775. [CrossRef]
11. Xue, L.; Shen, Z.; Boroyevich, D.; Mattavelli, P.; Diaz, D. Dual Active Bridge-Based Battery Charger for Plug-in Hybrid Electric Vehicle with Charging Current Containing Low Frequency Ripple. *IEEE Trans. Power Electron.* **2015**, *30*, 7299–7307. [CrossRef]
12. De Caro, S.; Testa, A.; Triolo, D.; Cacciato, M.; Consoli, A. Low input current ripple converters for fuel cell power units. In Proceedings of the 2005 European Conference on Power Electronics and Applications, Dresden, Germany, 11–14 September 2005.
13. De Melo, H.N.; Trovao, J.P.F.; Pereirinha, P.G.; Jorge, H.M.; Antunes, C.H. A Controllable Bidirectional Battery Charger for Electric Vehicles with Vehicle-to-Grid Capability. *IEEE Trans. Veh. Technol.* **2017**, *67*, 114–123. [CrossRef]
14. Restrepo, M.; Morris, J.; Kazerani, M.; Canizares, C. Modeling and Testing of a Bidirectional Smart Charger for Distribution System EV Integration. *IEEE Trans. Smart Grid* **2016**, *9*, 152–162. [CrossRef]
15. Raciti, A.; Musumeci, S.; Chimento, F.; Privitera, G. A new thermal model for power MOSFET devices accounting for the behavior in unclamped inductive switching. *Microelectron. Reliab.* **2016**, *58*, 3–11. [CrossRef]
16. Mohammed, S.S.; Devaraj, D. Simulation and analysis of stand-alone photovoltaic system with boost converter using MATLAB/Simulink. In Proceedings of the 2014 International Conference on Circuits, Power and Computing Technologies [ICCPCT-2014], Nagercoil, India, 20–21 March 2014; pp. 814–821.
17. Doolla, S.; Bhat, S.S.; Bhatti, T.; Veerachary, M. A GUI based simulation of power electronic converters and reactive power compensators using MATLAB/SIMULINK. In Proceedings of the 2004 International Conference on Power System Technology, The Pan Pacific, Singapore, 21–24 November 2004.
18. Aiello, G.; Cacciato, M.; Scarcella, G.; Scelba, G.; Gennaro, F.; Aiello, N. RealTime emulation of a three-phase vienna rectifier with unity power factor operations. In Proceedings of the 2018 ELEKTRO, Mikulov, Czech Republic, 21–23 May 2018.
19. Aiello, G.; Scelba, G.; Scarcella, G.; Cacciato, M.; Tornello, L.; Palmieri, A.; Vanelli, E.; Pernaci, C.; Di Dio, R. Real-Time Emulation of Induction Machines for Hardware in the Loop Applications. In Proceedings of the 2018 International Symposium on Power Electronics, Electrical Drives, Automation and Motion (SPEEDAM), Amalfi, Italy, 20–22 June 2018.
20. Aiello, G.; Tornello, L.D.; Scelba, G.; Scarcella, G.; Cacciato, M.; Palmieri, A.; Vanelli, E.; Pernaci, C.; Di Dio, R. FPGA-Based Design and Implementation of a Real Time Simulator of Switched Reluctance Motor Drives. In Proceedings of the 2019 21st European Conference on Power Electronics and Applications (EPE '19 ECCE Europe), Genova, Italy, 2–5 September 2019.
21. Scelba, G.; Scarcella, G.; Cacciato, M.; Aiello, G. Hardware in the loop for failure analysis in AC motor drives. In Proceedings of the 2016 ELEKTRO, Strbske Pleso, Slovakia, 16–18 May 2016.
22. Acquaviva, A.; Thiringer, T. Energy efficiency of a SiC MOSFET propulsion inverter accounting for the MOSFET's reverse conduction and the blanking time. In Proceedings of the 2017 19th European Conference on Power Electronics and Applications (EPE'17 ECCE Europe), Warsaw, Poland, 11–14 September 2017.
23. Gritti, G.; Adragna, C.; Industrial & Power Conversion Division Application Laboratory—STMicroelectronics s.r.l.—via C. Olivetti 2—20864 Agrate Brianza (MB)—Italy. Analysis, design and performance evaluation of an LED driver with unity power factor and constant-current primary sensing regulation. *AIMS Energy* **2019**, *7*, 579–599. [CrossRef]
24. Chimento, F.; Raciti, A.; Cannone, A.; Musumeci, S.; Gaito, A. Parallel connection of super-junction MOSFETs in a PFC application. In Proceedings of the 2009 IEEE Energy Conversion Congress and Exposition, San Jose, CA, USA, 20–24 September 2009; pp. 3776–3783.
25. Yazdani, A.; Iravani, R. *Voltage-Sourced Converters in Power Systems: Modeling, Control, and Applications*; John Wiley & Sons: Hoboken, NJ, USA, 2010.
26. Cacciato, M.; Scarcella, G.; Scelba, G.; Finocchiaro, L. Multi-reference frame based PLL for single phase systems in voltage distorted grids. In Proceedings of the 2014 16th European Conference on Power Electronics and Applications, EPE-ECCE Europe 2014, Lappeenranta, Finland, 26–28 August 2014.
27. Jafari, M.; Malekjamshidi, Z.; Zhu, J.G. Analysis of operation modes and limitations of dual active bridge phase shift converter. In Proceedings of the 2015 IEEE 11th International Conference on Power Electronics and Drive Systems, Sydney, Australia, 9–12 June 2015.

28. Cacciato, M.; Consoli, A. New regenerative active snubber circuit for ZVS phase shift Full Bridge converter. In Proceedings of the 2011 Twenty-Sixth Annual IEEE Applied Power Electronics Conference and Exposition (APEC), Fort Worth, TX, USA, 6–11 March 2011.
29. Alves Ferreira Júnior, A.; Justino-Ribeiro, J.; Ney do Amaral Pereira, W. Evaluating Impedance Transformers with a VNA. *Microwaves Rf* **2009**, *48*, 64–71. Available online: https://www.researchgate.net/publication/243055235_Evaluating_Impedance_Transformers_With_a_VNA/references (accessed on 10 October 2020).
30. Barrios, E.L.; Ursua, A.; Marroyo, L.; Sanchis, P. Analytical Design Methodology for Litz-Wired High-Frequency Power Transformers. *IEEE Trans. Ind. Electron.* **2015**, *62*, 2103–2113. [CrossRef]
31. Hoang, K.D.; Wang, J. Design optimization of high frequency transformer for dual active bridge DC-DC converter. In Proceedings of the 2012 XXth International Conference on Electrical Machines, Marseille, France, 2–5 September 2012.

Publisher's Note: MDPI stays neutral with regard to jurisdictional claims in published maps and institutional affiliations.

MDPI

St. Alban-Anlage 66

4052 Basel

Switzerland

Tel. +41 61 683 77 34

Fax +41 61 302 89 18

www.mdpi.com

Energies Editorial Office

E-mail: energies@mdpi.com

www.mdpi.com/journal/energies